Circuits in the Brain

Charles R. Legéndy

Circuits in the Brain

A Model of Shape Processing in the Primary Visual Cortex

Springer

Charles R. Legéndy
Psychology Department
Columbia University
Schermerhorn Hall
1190 Amsterdam Avenue
New York, NY 10027
USA
crl2122@columbia.edu

ISBN 978-0-387-88848-4 e-ISBN 978-0-387-88849-1
DOI 10.1007/978-0-387-88849-1

Library of Congress Control Number: 2008942079

Printed on acid-free paper

springer.com

To Annemarie, Philip, Gabriel, Conrad, and Gwendolyn

Preface

The idea that circuits might be able to explain brain function never really occurred to me until once when, as a Princeton undergraduate, I heard a lecture about a Cornell psychologist, Frank Rosenblatt, and the machine he had designed, the *Percepron*, which could "learn" to recognize letters of the alphabet. The lecture planted a seed, but it took very long before anything grew from it.

I went to Cornell after graduating from Princeton, to work on a PhD in theoretical physics, and made occasional visits to Rosenblatt's lab. I kept telling him how I didn't think the brain worked the way he described, and he countered by pointing to the large number of "theories" I could choose from. Of course I had no idea how the brain worked, but in the process I started reading books on it.

It was hard to know where to start. Physiology, anatomy, psychology, neurology, computer science – these were all subjects directly relevant to what I needed to know, not to mention that data and facts about the human mind are all around us; and observations too common to require laboratory research are not automatically unimportant.

I finished my PhD work (on solid-state plasma oscillations known as *helicons*), and proceeded to a job with United Aircraft (now United Technologies) in Hartford, as a physicist under a contract which permitted me to spend one third of my time on work I chose, which in my case would be brain theory.

From Hartford I often drove up to MIT, where McCulloch had his office, cluttered with papers, and Lettvin his lab, with the name tag "J. Y. Lettvin, experimental epistemology." They were both still active, and gave generously of their time. At United Aircraft I wrote my first brain theory paper, on ignitable neuron groups (Legéndy, 1967). Influenced by the approach of McCulloch, Rosenblatt, and von Neumann, the 1967 paper attempted to explore the duality between subjective perception and the facts of connectivity, and describe a set of "networks with circles" which are more reliable than their individual elements.

The proper career path for me at this point would probably have been to enroll in a graduate school again and study under some established neuroscientist, but by this time I have become aware that there was nobody able to teach me what I needed to know. I was navigating uncharted waters.

My good luck steered me to a small Westinghouse research group, located right on "Tech Square," a block of research buildings across the street from MIT, where

I was offered a job doing pure research in brain theory, justified as potentially leading to computing machines better suited to pattern recognition than digital computers. Tech Square was the world's greatest place to learn about brains. I often spoke to McCulloch, got to know Minsky and Papert, and, on the other side of the river, at Harvard Medical School, got to visit the labs of Hubel, Wiesel, and Palay.

While at Westinghouse, I wrote down many of the principles which underlie the present book; among them were the need to formulate communication inside the brain in terms of "surprising" events of firing, the conceptual linkage between them and "local knowledge," and the idea that neuron groups representing objects can transmit "syntactic" relations between the objects through prearranged relative timing of their outputs (Legéndy, 1970). I also wrote down the idea of "reaching," and some methods the brain uses to achieve it, such as the "trick of retinotopic mapping" and the "trick of small connective fields" (Remarks on the Brain, unpublished internal report, 1970).

After my Westinghouse money ran out, I took a postdoctoral position in Italy, at the lab of Caianiello (at the CNR Laboratory of Cybernetics at Arco Felice), then another one in Germany (at the University of Tübingen), a few minutes away from the Max Planck Institute of Biological Cybernetics, where Valentino Braitenberg had his group. I spent many good hours at Braitenberg's office looking at Golgi slides through his microscope.

I still had much to learn. For one thing, I had felt since my conversations with Hubel that my education would remain incomplete until learned to do experiments of my own. My years in electrical engineering (which I had studied at Princeton) steered me toward electrophysiology, and an opportunity to learn the techniques soon arose at Otto Creutzfeldt's lab in Göttingen, where I had a chance to join a project with Tadaharu Tsumoto on cortico-geniculate correlations. At Göttingen I went on to spend a very busy one and a half years, and wrote, among other things, a program to make "Poisson surprise scans" of some spike trains I recorded. By the time I returned to the New York City, I knew enough to take postdoctoral jobs as a neurophysiologist, first with Alden Spencer at Columbia University, then with Herb Vaughan at Albert Einstein College of Medicine.

But my career in experimental brain research was short-lived, because after a while I lost my NIH funding and had to take jobs in the aerospace and computer industry. It seemed that my brain research days had come to an end. Except that around this time my good luck once again intervened.

Wayne Wicklegren, formerly of MIT and the University of Oregon, moved to New York and joined his wife Norma Graham at Columbia University. I knew both Wayne and Norma from the literature, as they both knew me; in particular Wayne, encouragingly, had a lively interest in my 1967 paper. We talked about the brain on many afternoons for several years; and by the time we stopped, I was doing brain research once again, this time not experiments but theory. Along with my wife, Annemarie, who patiently stuck it out with me throughout, Wayne deserves my deepest thanks for the push he gave me when I most needed it.

During working hours I continued my computer jobs, but before and after work I started filling notebooks with my thoughts on vision; then after retiring I started typing them into a computer, until this book came together.

The book you see here does not deal with the whole brain, only with vision, and within that subject only with one class of large image-determined circuits: the *contour strings*. My original intent was to include another class of large circuits, the "color pools," and I had also hoped to extend the discussion beyond the V1 area of visual cortex, to V2, where the first version of a "stable" cortical image (which does not move around with the retinal image) promises to arise.

However, preliminary work convinced me that both of the other subjects, the *color pools* and the *stable image*, were hugely more complex than the *contour strings* and I decided to leave them out, at least for now.

New York, NY Charles R. Legéndy

Contents

List of Figures

Introduction

If one had to name the single most noticeable thing about the brain, one would, without a doubt, say that it was the rich branching and connectivity of the neurons. There are kilometers of fiber per cubic millimeter and thousands of synapses per neuron, and they clearly convey the impression that the brain must be good at making connections.

In fact it is legitimate to argue that one function of the brain is to translate "connections" in the metaphorical sense into *connections* in the physical sense; in other words, to transform connections as exist between things of the world, when they are linked up in our mind, into connections which are tangible, and take the form of fibers and synapses.

In this book I will propose some examples of such translation within the brain; even though, upon first approach to the subject, the two kinds of "connection" do not appear to translate into one another in any obvious way. It will not do, for instance, to describe the creation of connections as an act of lumping mental images into unstructured bagfuls of details. The connection between the things tied together by our mind is presumably some form of *syntactic* linkage; a cup of coffee is more than just "cup; coffee." The challenge being addressed is to understand how the neurons and synapses, together with some rules, manage to accomplish syntactic linkage.

The problem is closely related to what, in a more general formulation, has become known as the *binding problem* (von der Malsburg, 1981, 1999), and the following monograph may be considered an attempt to solve the binding problem for a very limited case: the processing of individual contours in the primary visual cortex. Although the present model does not arrive at the complex syntax of spoken language (the kind linking *cups* to *coffee*), it relies in essential ways on a rudimentary form of syntax.

Persons familiar with the *cell assembly* approach to brain modeling (Hebb, 1949; Legéndy, 1967; Palm, 1982; Wickelgren, 1999) will recognize many of its usual elements in the model described below, but often with a twist. There will be cell assemblies, but they will not be permanent but created on the fly, under the influence of visual input (since the model deals with vision), and will continually change their membership as the retinal image drifts, then disappear at the end of a fixation.

They will be capable of *ignition* (Legéndy, 1967; Braitenberg, 1978; Palm, 1982), but in the model below each ignition will consist of only a single spike emitted by each neuron of the assembly; the ignition will not be followed by continued reverberation. Accordingly, the claim will be made that many of the neuronal spikes seen in single-unit recording, especially the extra spikes during certain *bursts* of firing (Legéndy and Salcman, 1985), are actually ignitions, indistinguishable from solitary spikes in single-cell records.

The present book is published at a time of rapid advancement in experimental knowledge, fueled by remarkable technical innovations, some of them changing the way synaptic plasticity and other fundamental brain-related matters are perceived. Although the outlines of the original Hebbian (1949) synapse modification rule survive, the rule must, as of recently, be changed to permit synapse modification in the absence of postsynaptic firing (Remy and Spruston, 2007). Other changes can also be expected, describing, for instance, mechanisms for varying the cellular parameters when needed.

Over the last years, the concept of recording from neurons has taken on a wholly different meaning. The new literature includes dual whole-cell recording from synaptically linked neurons (Markram et al., 1997); voltage-sensitive and ion-sensitive dyes making dendrites light up as they respond to input (Djurisic et al., 2004; Baker et al., 2005); two-photon microscopy permitting the visualization of single synapses separated from the microscope's objective by hundreds of microns of brain tissue (Denk et al., 1990), which opened the way to documentation of the appearance and disappearance of synapses over time (Stettler et al., 2006); and the programmed stimulation of preselected sets of synapses by sub-micron-accurate laser uncaging of glutamate (Losonczy and Magee, 2006). In addition, the techniques of extracellular recording have now matured to permit massively parallel recording of cells from cortical tissue (Buzsáki, 2004).

The understanding of nerve cell chemistry has also undergone stunning changes. In recent years, more than a hundred new molecules, including receptors, messengers, scaffold proteins, protein kinases, have been identified as participating in synaptic function (Muller et al., 2008).

Arguably, it would be wisest for me to wait 10 years before publishing this book, as it contains some predictions sure to be disproven in the next few years. However, alternatively, it can also be said, and it is the viewpoint I prefer, that the book is published at an opportune time, because it permits placing the facts, as they are discovered, side by side with certain non-obvious functional requirements revealed in the book.

The approach, to be described in more detail below, starts by saying that the brain does not operate through trial and error. The duration of the average life (or for that matter the age of the universe) would be too short to overcome the combinatorial explosion if learning were to rely blindly on trial and error at every locality large and small. Which is the same thing as saying that no piece of tissue undergoes change without a good reason, and that the tissue cannot know that it must change if the necessary information has no way of getting to it.

Formal theoretical research flows within fixed confines; classically, the confines are provided by mathematics, in more recent work they are provided by that and computer simulation (Beurle 1956; Arbib 1964, 2003; von der Malsburg 1973;

Wilson and Cowan, 1972). All of it is meant to keep the theory fixed within its own constraints of correct thought. In the present book I aim to demonstrate the effectiveness of another form of constraint-based reasoning: the one just mentioned, which says in essence that all information must be able to reach the localities where it is needed (usually in the form of coincident firing). It, too, is rigorous and leads to narrowly defined possibilities, although (in the form used here) it depends neither on equations nor on computer programs. The presentation given below is entirely non-mathematical.

It may be mentioned as a general observation that non-mathematical description like the one in this book, featuring mainly annotated drawings and text, is not new to engineering or the patent literature. It is often the method of choice when it comes to unambiguous functional description of complex machines whose current form is the result of a sequence of inventions, such as the printing press or the manual transmission. I propose, it is also the method of choice when it comes to describing the brain.

In the detailed sequences shown in this book, the description usually extends no further than to neurons and column-sized localities. Speculations concerning sub-cellular mechanisms will be replaced by the most optimistic assumption possible, which is in this case the assumption that whatever these elements need to do in order to support the overall function they are biologically equipped to do. The result produces a kind of "theoretical skeleton" of the brain, delineating what its pieces "can" do, provided they are equipped to do it.

The optimistic mode of reasoning, viewed in another way, amounts to a top-down parsing of overall brain function, whereby known features of the system are broken down to tasks addressed by subsystems, and those to tasks addressed by smaller subsystems, eventually reaching the atoms and molecules. The parsing breaks up the problem of the brain, too large in its entirety to be addressed in a single publication, into smaller pieces.

The restriction that no piece of tissue can act on information unable get to it reduces some of the modeling issues, as will be seen, to what amount to problems of *logistics*, the art of arranging that all necessary materials are available wherever needed and whenever needed. The difference between the present version of logistics and the traditional version is that in the present case the materials required are pieces of "local knowledge." Such logistic analysis has been found to be very effective in limiting the choice of possible protocols available at the network level (see for instance the sequences in Sects. 19.1 and 20.1).

In higher animals the majority of contacts between neurons are by necessity unplanned, except for specifications that direct certain neuron pools to contact certain other neuron pools; accordingly the synapse-for-synapse description of meaningful input patterns is often not a priori known to target cells. If the multi-channel stream of signals arriving to neurons and neuron groups is to impart the local knowledge needed, it is necessary that the locally arriving firing be recognized, from one occurrence to the next, in a way which is correct with very great likelihood.

This in turn means that the multi-channel firing events causing local change must be "surprising" (Legéndy, 1970, 1975; Palm, 1981; Legéndy and Salcman, 1985), which is approximately the same as saying that the evidence of their distinction from background noise must be statistically significant.

The model presented below will address the ability of the visual cortex to pass the building blocks of *shape information* to the higher cortex, rather than passing only the individual receptive field responses. Much attention will be paid to the ability of vision to deal with the drifting of the retinal image, and an explanation will be offered for the fact that the drifting is even *helpful* to vision.

The present model must by necessity deal with the fact that, at the current stage of research, some crucial experimental evidence is missing at almost all levels of the system. Where this is the case, the approach chosen is to offer a reasonable level of certainty that with the right choice of parameters the described sequences will work.

The following is a brief summary of the book:

Part I. Concepts of brain theory. This part introduces the first few components of the present approach (the rest are introduced later, as needed), then presents a demonstration of the way the finished model can convey shapes. Signals arriving to neurons are described in terms of multineuronal "events," noticeable (*surprising*) enough to stand out above the background noise. Of special interest are the events generated by groups of interconnected neurons (*cell assemblies*) within which firing can spread until each neuron in a group emits a spike (*ignition*). Reverberation within the groups is prevented by the relatively long recovery time of the neurons, which only permits each neuron to emit one spike per ignition with the result that each ignition gives rise to a *volley* of nearly simultaneous spikes. The visual environment is by its nature organized, but the organization does not automatically give rise to surprising events inside the brain. Such events only arise if the network is designed to generate them; much of the rest of the book deals with the tricks of design able to achieve that. In way of linking the idea of surprising events to shape perception, it is shown that sets of simultaneous ignitions by pairs of cell assemblies (which are simple visual *sentences*) can be joined together (on their shared *noun* objects), with the result that many two-element relations are combined into single multi-element relations, and used to broadcast whole complex shapes.

Part II. Contour strings and the contour wave. The next part describes a kind of temporary cell grouping (*contour string*) which arises whenever a contour appears in the underlying retinal image. *Simple*-type cells with adjacent and similar receptive fields join hands along every contour, in a way that enables them to pass spikes in both directions along the contour, and spontaneously initiate such propagating spikes (*contour waves*) as long as the contour is present in their receptive fields. It is necessary to address the issue that the retinal image of the contour does not hold still even during periods of fixation. As a result, the simple cells must be enabled to pass waves immediately when the contour enters their receptive fields, which they can do only if they can be "warmed up" with the help of nearby simple cells, before the contour arrives to them. The role of the *complex cell* is described in connection with a need for cells outside the contour string to be able to follow (*track*) the drifting contour wave. A sharp turn in the contour would cause contour waves to die if they relied on ordinary simple cells alone, because similarity of preferred orientation is utilized in making the cell inputs surprising enough to pass the wave. Accordingly, uninterrupted propagation around corners requires a pool of special cells able to pick up the contour waves from nearby cells with greatly differing preferred orientation (*corner-supporting simple cells*, which are hard to distinguish from the *hypercomplex* cells) and their corresponding tracking-enabling companion cells (the actual *hypercomplex* cells).

Part III. Nodes, links, bridgeheads. This part of the discussion addresses several practical considerations involving the grouping of neurons. One is that all ignitions must involve enough neurons to impart a noticeable volley on their target neurons; another is that the retinal image, by continually drifting, causes the neuron groups along the contour (*nodes*) to keep reconstituting themselves, and continually dropping neurons while recruiting new ones. One node is able to link up to a number of other nodes by hosting a separate ignitable neuron group (*bridgehead*) for each of the links it makes. To make tracking and the continual reconstitution of groups possible, the bridgeheads on interlinked nodes, once created, repeatedly go into instability (*self-ignition*) on a random schedule, and provide a steady flow of status reports on the linkage. As the contour drifts, the neurons which no longer receive receptive field stimulation are not automatically dropped from the igniting groups but are kept on until *echolocation* (made possible by the repeated self-ignitions) tells them that mutual contact exists. As a result, thanks to contour drift, bridgeheads which are too far apart, and at first do not have enough long-axon neurons to transmit detectable volleys to the other node, have a chance to grow until they do.

Part IV. Firing games and the integration of contours. To transmit the shape of a contour, the pairs of nodes which are to form active links must be on the same contour, and accordingly the first challenge after creation of the contour strings, and creation of the nodes on them, is to use the contour string as an avenue on which to move from one node until encountering the next node which is then, automatically, on the same contour. The next task is to set up two-way communication between the nodes in preparation for linkup. It is necessary to assemble sequences of local decisions (Sects. 19.1 and 20.1) resulting in linkup and repeated co-ignitions. Each sequence (*firing game*) requires coordinated multi-neuronal changes which proceed without any neuron playing the role of a leader. In Chapters 19, 20, 21, and 22, the linkup of nodes on a contour is described in two phases: in the first, neighboring nodes are linked up, and in the second the existing links are utilized to establish further links. It will be appreciated that, before being linked, the node pairs are only connected "in the metaphorical sense," since the only thing initially connecting them is the fact that the image of some contour falls upon both of them.

Part I
Concepts of Brain Theory

Chapter 1
Lettvin's Challenge

A good starting point in a discussion of shape perception is the paper of Lettvin, et al. (1959), who recorded from the frog's optic nerve fibers and classified the retinal ganglion cells, from which the fibers originate, with regard to their responses to visual stimulation. On the basis of these observations the authors divided the cells into four types: *sustained contrast detectors*, *net convexity detectors*, *moving edge detectors*, and *net dimming detectors*. From the descriptions it was clear that the outputs of many of these cells were more elaborate than the outputs of the corresponding cells in higher animals, but still, all told, the four cell types were not much more than detectors of moving edges and dots, or changes of illumination.

Lettvin gave his paper the challenging title "*What the frog's eye tells the frog's brain,*" calling the professional community's attention to the fact that since the optic nerve is interposed between the eye and the brain, "in series" as it were, it appears from the finding that nothing more can get to the brain from the eye than some edges, dots, and whatnot. No compound shapes, for instance.

The professional community took it calmly. People assumed, perhaps, that a primitive creature like the frog did not need to see shapes to stay alive, only edges and dots. However, around the same time, Hubel and Wiesel (1959) recorded from the cat visual cortex, a structure also interposed in series somewhere between the eye to the rest of the brain, and showed that by the same argument cats couldn't see anything but edges and bars; then a few years later (Hubel and Wiesel, 1968), that monkeys couldn't either.

The papers of Hubel and Wiesel held some promise of progress toward the synthesis of complicated patterns, because they demonstrated, in mammals at least, a hierarchical progression of receptive field properties, from retinal ganglion cells to cortical simple cells to cortical complex cells and hypercomplex cells. The effort to locate cells responding to complex images (Gross et al., 1972), like the elusive "grandmother cell," the hypothetical cell responding to a particular face and no other, still receives occasional attention (Sheinberg and Logothetis, 2001, Gross, 2002).

The effort to understand the anatomical and physiological facts that underlie shape perception and the *integration of contours* has been substantially advanced by the successful staining of functionally identified cells of the primary visual cortex with long horizontal axons, and the discovery of correlated firing between such cells,

C.R. Legéndy, *Circuits in the Brain*, DOI 10.1007/978-0-387-88849-1_1,

some of them several millimeters apart in the visual cortex (Gilbert and Wiesel, 1979, 1983). In order to characterize the non-local inputs to these cells, the classical receptive field concept has been extended to include the study of subthreshold effects from secondary parts of the receptive field which map, in correspondence with the other studies, to the cortical columns several millimeters away from the columns containing the cell bodies (Ts'o, Gilbert, and Wiesel, 1986; Gilbert and Wiesel, 1990; Seriès et al., 2003).

Theoretical study of the same questions has been addressed, formulated mainly in terms of average spike rates (Zhaoping Li, 1998), but the concrete steps of shape processing have still not been addressed. It is clear that our visual system can perceive complex patterns, and also that the primary visual cortex has a way to set up the data in a form that preserves shape information, but to my knowledge the present monograph is the first systematic attempt to describe the manner in which it can do so.

The premise of the approach described here is that the visual cortex, instead of permanently "welding together" a set of features to make shape detectors (as is implicit in the *grandmother neuron* concept), ties them together temporarily for dynamic transmission. In effect, it custom-designs a network for the transmission of each shape as it arises.

Chapter 2
Issues Concerning the Nature of Neuronal Response

2.1 Impressions Gained from Histograms and Raster Displays

A person whose intuition of neuronal reliability is shaped by post-stimulus time (PST) histograms (Gerstein and Kiang, 1960) and dot raster displays (see for example Schmidt et al, 1975) (myself included, in the early papers Legéndy, 1970, 1975) may well be tempted to assume that the single neuron is only reliable in broad statistical terms. To elicit reproducible behavior from a neuron of the visual cortex, for instance, one must sweep the receptive field 10–20 times, combine the sweeps into a PST histogram or a raster display, and examine the way the spikes are distributed. Two or three sweeps do not appear to be enough to get a clear idea of the neuron's behavior, because, as seen in Fig. 2.1, the responses do not repeat spike-for-spike; some sweeps show more spikes, some fewer.

The displays suggest to us that a neuron responds, when it does, by emitting *a number of* spikes. Some typical responses are "brisk," where a number of spikes,

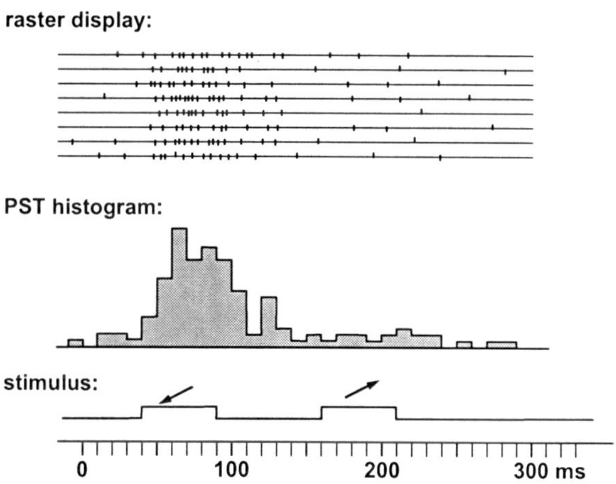

Fig. 2.1 Raster display and post-stimulus time histogram. *Top*: responses of a neuron to eight triggered sweeps of a stimulus (hypothetical neuron and stimulus); *bottom*: spike counts gathered from the same spike trains (10 ms bins)

C.R. Legéndy, *Circuits in the Brain*, DOI 10.1007/978-0-387-88849-1_2,

say 5–10, are crowded into a short time interval, say 20–30 ms; others are more "sluggish" where fewer spikes appear and they are spread over a longer time interval. They do not encourage us to trust the neuron to emit spikes individually timed in ways that fit into the overall processing.

However, recent studies tell us that neuronal outputs are quite reliable, as long as the inputs are. The internal signal conduction within the neuron has been shown to be both reliable and fast (Mainen and Sejnowski, 1995; Ariav et al., 2003), and in particular much faster than had been believed on the basis of the classical resistive-capacitive picture of dendrites (Rall, 1962).

The reliability and speed of neurons seemingly contradicts their irregular mode of firing in behaving animals, and leads to a useful inference concerning the neuronal input stream, as will be seen next.

2.2 Cortical Firing Should Be Nearly Periodic – So Why Isn't It?

When a steady current is injected into the interior of a neuron (Fig. 2.2), the neuron emits more or less periodic firing, whose spike rate increases with the current; it does not for instance fire irregularly at a current-dependent average rate. More recently it has been demonstrated (in vitro) that current injection into apical dendrites, and even thin (1.9 micron) basal dendrites, or comparably thin branches of the dendritic tuft in layer 2–3 pyramidal cells, causes similar and more or less regular firing (Larkum et al., 2001; Nevian et al., 2007; Larkum et al., 2007).

The counterintuitive aspect of the observation becomes clear when one notes that, at least in a crude approximation, every spike coming to an excitatory synapse injects a small amount of electric charge into the cell interior. Since typically thousands of spike trains arrive to a cortical neuron through its synapses, their steady rainfall is expected to simulate a DC current injected into the cell interior, and cause more or less steady firing like what is seen in Fig. 2.2, or at least steady firing at a slowly varying rate, as shown in Fig. 2.3(a). However, in fact, typical spike trains in the visual cortex look something like Fig. 2.3(b).

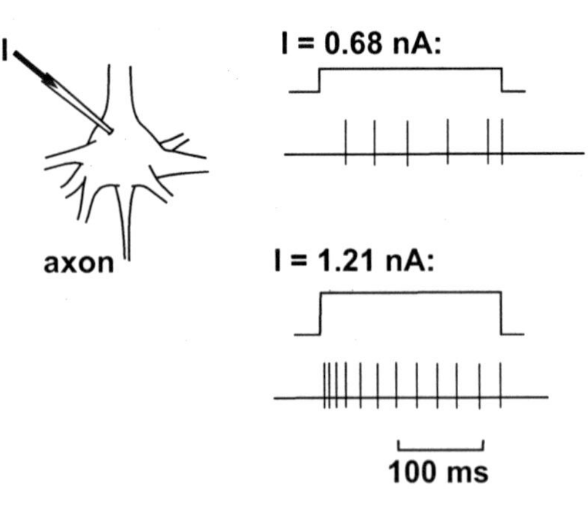

Fig. 2.2 Response of a cell to intracellular injection of DC current. Response (in vivo) of a Betz cell from the motor cortex, when a pulse of steady electrical current, lasting about 200 ms, is injected into its cell body (Anesthetized cat. Re-drawn from Creutzfeldt et al., 1964)

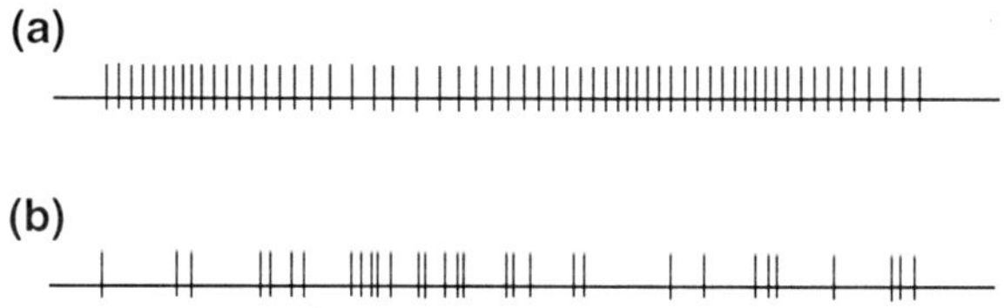

Fig. 2.3 Effect of input correlation on the appearance of neuronal output. (**a**) Appearance of the firing of a cortical neuron on the assumption that its input spike trains are only crudely correlated (to an accuracy of about hundred ms). (**b**) Appearance of firing as seen in a typical recording from the visual cortex of an unanaesthetized cat

The contradiction was first raised in the form of a problem requiring a solution by Perkel and Bullock (1969) who used the mathematical results of Cox and Smith (1954); it was later brought to wider attention by Softky and Koch (1993), and has since then given rise to a fair amount of discussion in the literature. It may be remarked that some of this discussion incorrectly implies that current injection only causes regular spiking in brain slices (in vitro), whereas in fact it also does so in the intact brain (Creutzfeldt et al., 1964; Oshima, 1969; Ahmed et al., 1993).

The irregularity of firing in response to natural inputs has been attributed by some writers to brief pauses in firing caused by volleys of inhibition (Shadlen and Newsome, 1998), and by others to noise in the membrane and synapses (Destexhe et al., 2001). Both groups of authors see the irregular firing as resulting from noise-like random events of one sort or another.

The explanation in terms of random events has never been entirely satisfying. Ahmed et al. (1993), in an in vivo study, point out (and it is also observable in the data of Creutzfeldt et al., 1964) that at low levels of injected current the firing is often irregular, but at stronger currents it becomes more regular. The random noise explanation would predict that during the brief epochs of elevated spike rate (bursts) reported in waking animals (Legéndy and Salcman, 1985), when electric charge is expected to enter the cell interior at an increased rate, the irregularity-causing effects of noise would tend to be overwhelmed by the input current, and the spike discharge would become more uniform (Holt et al., 1996).

However, in behaving animals the spiking is just as irregular during the bursts as it is outside the bursts (Legéndy and Salcman, 1985). In must be added that Stevens and Zador (1998), in a careful statistical study, eliminate both noise and inhibition as reasons for the observed irregularity, and show that their effect is, at least in an in vitro preparation, insufficient for causing the observed behavior. While the role of inhibition in causing the irregular firing is likely to be important, the conclusion remains that the irregular firing of the neurons in behaving animals implies synchronized and correlated synaptic input.

The present model goes along with the latter conclusion and assumes that a significant number of spikes arriving to the neurons are sharply synchronized.

The idea of synchrony sharp enough to dictate the timing of output spikes suggests that the incoming spikes arrange a "rendezvous," in some way, on the receptive surfaces of neurons, which is a seemingly outlandish idea and may account for some of the resistance to it in the literature.

As will be seen, the concept becomes much more plausible when viewed from a different perspective (Sect. 4.2), but before getting to that, let me make a brief remark on the effect of volleys of spikes on neurons, and then survey some of the literature on plasticity.

2.3 Sensitivity of Neurons to Synchronized Volleys of Spikes

The form of "rendezvous" which comes up most often in this book is one which sends a volley of nearly simultaneous spikes to neurons, along the lines shown in Fig. 2.4. Such volleys clearly stand out above the background noise, especially when

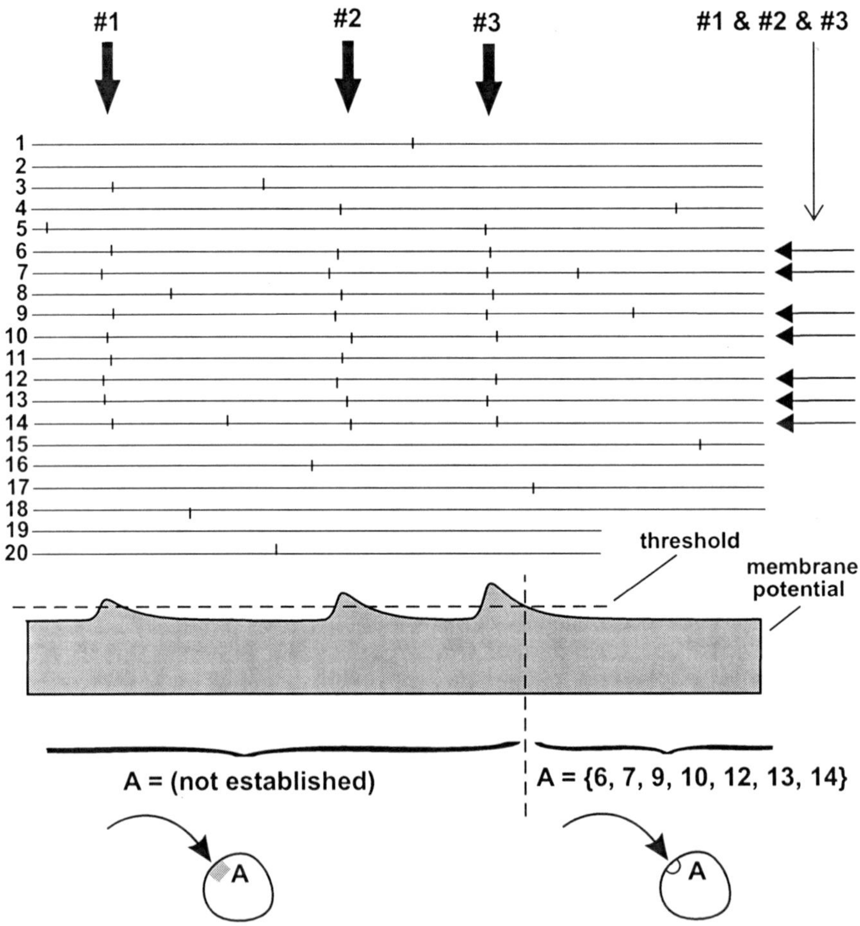

Fig. 2.4 (continued)

repeated, and there is evidence that if they arrive repeatedly they can cause plastic change, marking the synapses on which they arrive.

In Fig. 2.4, repeated volleys arrive to a set of synapses (arrows at right), envisioned as being located sufficiently close together on a dendrite to cause the indicated excitatory postsynaptic potentials (EPSPs) at the synapses. The gradually increasing EPSP amplitude (exaggerated) is intended to imply a mechanism of short-term and long-term potentiation (STP and LTP). The *threshold* marked with broken line stands for the minimum level of membrane potential needed, in conjunction with presynaptic activity, for inducing synaptic potentiation.

The insets at the bottom of Fig. 2.4 introduce a graphical notation (see also Sect. 12.3), to be used in the functional diagrams of later chapters for sets of synapses on neurons which undergo LTP, or are "marked," as will be said from now on. A set of synapses, initially unassigned and only known to be a subset of the synapses from a biologically distinct neuron pool (rectangular shading at left), becomes "marked" (semicircle at right) by repeatedly arriving volleys. The amorphously drawn blob shape in these drawings stands for the typical neuron from among a *set* of similar neurons, or alternatively for the neuron set itself, and the arrow for a set of fibers coming to the neuron and bringing the signals which do the marking.

2.4 Notes on Plastic Change at the Synaptic Level

The classic Hebbian scheme of synapse modification (Hebb, 1949) does not permit plastic change to occur as a result of time-concentrated presynaptic events alone; it requires the postsynaptic neuron also to fire, right after the presynaptic firing, and until recently all experimental findings agreed. It appeared that only through firing, and the resulting back-propagated action potential, could a neuron send all its participating synapses an intense enough and fast enough wave of depolarization to induce them to long-term change.

However, recently, Remy and Spruston (2007) showed that the postsynaptic firing is not strictly necessary, as long as the synaptic invasion is intense enough to give rise to *dendritic spikes*. It may be added that, in practice, when volleys are intense enough to cause dendritic spikes, they are also expected, after some repetitions, to cause action potentials, since the level of current injection required for the latter is not much greater (Nevian et al., 2007). (For the purposes of their demonstration,

Fig. 2.4 (continued) Marking a set of synapses participating in repeated volleys. Traces 1–20 stand for spike trains arriving to some of the synapses on a (hypothetical) neuron. At times #1, #2, #3, (*arrows on top*) the spike trains form "volleys"; seven of the spike trains (*arrows at right*) participate in all three of the volleys. The corresponding membrane potential (idealized) is shown below the traces. *Bottom*: synapse set notation symbolizing a whole synapse pool (A) on the neuron receiving input (*left, shaded rectangle*), and a "synapse set" marked out from among members of the same synapse pool as a result of the recurring volleys (*right, semicircle*)

Remy and Spruston, 2007, suppressed the action potentials and back-propagation by means of tetrodotoxin.)

According to the available evidence, induction of LTP requires presynaptic spikes in conjunction with either back-propagated action potentials (Magee and Johnston, 1997) or locally induced dendritic spikes (Remy and Spruston, 2007). Glutamate released in the synaptic transmission binds to the N-methyl D-aspartate (NMDA) receptors and opens up their calcium channels which are, because of the brief change in membrane potential, momentarily freed of the Mg2+ ions usually blocking them, and permit Ca2+ ions into the cell interior (Ascher and Nowak, 1988; Bliss and Collingridge, 1993).

It will be emphasized that Fig. 2.4 implies something not yet addressed by experimental data, namely that a set of synapses can be individually selected by the input arriving to them, even when they are not all adjacent; in other words, the synapses reinforced can be interspersed among synapses not reinforced.

From a functional point of view the capacity for such individual synapse selection is expected to be a requirement, because the arriving volleys come from cooperative neuron groups which, although functionally connected at the moment, are not similarly connected at the earlier time when axonal growth decides the location of their potential synapses. Accordingly they cannot be expected to have their synapses all located right next to each other.

The recently developed technique of two-photon glutamate uncaging at preselected sets of synaptic spines (Losonczy and Magee, 2006; Gasparini and Magee, 2006; Losonczy et al., 2008), now makes it feasible to address the issue of individual synapse selection. The results available at this point indicate that the synapses do not need to be concentrated on one portion of a dendrite but can be scattered throughout its length, but do not demonstrate the possibility that some synapses do and other nearby ones do not undergo potentiation. The issue is still open at the time of this writing.

Chapter 3
"Events" in the Brain

3.1 The Brain Viewed as a Logic Network Without a "System Clock"

The early classical papers of brain theory, specifically those of McCulloch and Pitts (1943) and John von Neumann (1956), made use of the concept that a neuron, which they treated as a threshold device (an approximation still considered valid), can be made to play the role of an AND gate or an OR gate in a logic circuit (and with a little acrobatics, an inverter), which means, by and large, that anything a digital computer can do, a network of idealized neurons can also do.

One subtle problem with these computer-like brain models, which treat the neurons as if they were gates in a logic circuit, is that implicitly they all require the brain to have a "system clock," the same kind the digital computer has, and for the same reason.

When sets of neurons perform neatly planned Boolean computations by combining their inputs to produce outputs, they must coordinate their outputs with the outputs of other neurons in order to meet at the next layer of neurons as they should; otherwise they would end up combined with the wrong signals.

So, in these papers, and in their immediate successors (Caianiello, 1961) which move away from logic gates to networks of threshold neurons, it is stated somewhere early that "time is quantized," or equivalently that time is "assumed to be an integer," which means that one can speak of the state of the network at "time n" or "time n+1."

It is noted that this assumption deals with the issue just mentioned, of arranging that neurons in the brain receive coordinated inputs. The quantized time acts like the system clock in computers and provides a natural solution to the issue, at least on paper.

The problem is that the brain does not have a system clock.

If it did, it would be experimentally verifiable with absolute certainty; the clock beats would show up in single-unit histograms, multi-unit records, evoked potentials, and the electroencephalogram. However, no evidence of any kind exists of an overall clock beat in any of these displays.

It is sometimes said that theories are easy to knock down, because a theory is rendered invalid if it makes even a single prediction which goes against the facts. This is of course wrong. Disagreements with experiment can often be "fixed up" by

C.R. Legéndy, *Circuits in the Brain*, DOI 10.1007/978-0-387-88849-1_3,

changing the assumptions in ways which leave the main framework of the theory intact.

Often but not always, the quantized-time-based models (and the spike-rate-based ones; Sect. 3.4) *cannot* be fixed up to square with the facts, such as Fig. 2.3 and the absence of an overall clock beat. These models must be re-written from the bottom up, and the rest of this monograph is an attempt to do that.

3.2 Looking for "Surprising Events" in the Neuronal Input Stream

Let us return briefly to Fig. 2.4, which shows volleys of spikes arriving to a neuron. From the perspective of the neuron, the volleys look like "events" within the multi-channel spike train, in the sense of the word familiar from probability theory, where one can speak of the *probabilities* of *events.*

In the present context, the events must be "surprising" if they are to have any effect on the neurons on which they impinge, meaning that they must have a low probability of occurring by accident. Otherwise the neurons, blind to the outside world and receiving nothing more than their synaptic input streams, have no way to tell the important inputs from the background.

The volleys shown in Fig. 2.4 stand out from the background by virtue of the brief surges they cause in the electrical current entering the cell, and there is no obvious reason to view their effect from the standpoint of probabilities. When the recurrences of the same volleys exert progressively greater influence on the cell, the probability-theoretic aspect of the recurring event becomes more apparent. The gradual potentiation of the participating synapses (assuming, as was mentioned, that the selective potentiation of synapses will be confirmed experimentally) can be considered a biological analog method of making the neurons sensitive to surprise.

In the rest of this monograph, the surprising nature of multi-synaptic events will be considered the item of primary importance; and for instance in Fig. 2.4 the bunched appearance of the spikes will only be regarded as a practical device designed to fit the biological surprise detector of the cells. It is considered likely that other surprise detection methods also exist in the nervous system. In Part II, in particular, another example will appear, where spikes generated by a moving wave arrive to recipient neurons in quick succession rather than in bunches (see for instance Fig. 16.1(b-e), bottom traces).

The theoretical importance of surprise arises from the shortage of a priori information concerning neuronal input patterns, forcing the decision-making elements to take a statistical approach to them.

As the brain develops early in life, the axons start growing from the cell bodies, find target neurons, and make synapses. It is considered likely that the broad neuron pools sought out by the growing axons are listed in the genetic blueprint. However, the visual system needs to be able to process information that cannot be anticipated in the genetic blueprint, and increasingly so in the higher animals. To process such information, the neurons must have the ability to detect finer groupings of input channels within the biologically determined pools and develop responses to them.

The "surprise" contained in an event was first quantified by Legéndy (1970, 1975), along the lines of *information* (Shannon, 1948), as the negative logarithm of a probability, in this case the probability of accidental occurrence. The expressions for *surprise* and *information* are comparable, but the two concepts are different, defined on different sets of events (Palm, 1981). In addition, surprising events bring zero surprise to a neuron until they are detected and the neuron is equipped for them, as in Fig. 2.4. In my 1970 paper, the word *surprise* was not yet introduced; instead the term "*meaningful information*" was used, as distinct from simple *information*.

3.3 *Poisson Surprise* as a Diagnostic Tool

The *surprise* concept was originally introduced in a theoretical context, but subsequently the "Poisson surprise" (so named because its baseline distribution for computing the probabilities was the Poisson distribution) was also used in the laboratory to identify bursts in the spike trains of spontaneously firing neurons (Legéndy and Salcman, 1985).

Since then, the idea found its way into diagnostic and other applications (see for instance Degos *et al.*, 2005, Starr *et al.*, 2005). Commercially purchased spike train analyzers equipped to do *Poisson surprise scans* are now sometimes used in clinical and research settings (for instance, Spike2, made by Cambridge Electronic Design, Ltd., Cambridge, England; and NeuroExplorer, made by Plexon, Inc., Dallas, TX, USA). The spike train scanning method from the 1985 paper was reformulated by Hanes *et al.* (1995), who subsequently made their code available on the Internet (http://www.psy.vanderbilt.edu/faculty/schall/code.html).

I may mention that the possible practical usefulness of the *Poisson surprise scan* first showed itself in the form of an "accidental discovery" of sorts, in the 1970s. I was a post-doctoral fellow at Prof. Otto Creutzfeldt's laboratory in Göttingen, recording from alert cats and trying to find out the way in which pattern vision would change the neuronal spike trains in the visual cortex. To do this, I compared records taken from cells in cats freely looking around in the lab with records taken from the same cells when the eyes were covered with translucent tissue paper and found that the difference was unimpressive. The tissue paper did not eliminate firing, or even bursts of firing; it only seemed to reduce the firing rate somewhat.

Then I scanned the spike trains with a new FORTRAN program I had written for detecting epochs of elevated spike rate, using the *surprise* idea I had developed a few years earlier, and found that the new program revealed a dramatic difference. The records taken with tissue paper showed no events with Poisson surprise beyond a fairly low value, but the records taken during free visual exploration showed a very large number of them, some with huge levels of Poisson surprise. (An explanation is offered below, in Sect. 9.9.)

After Salcman and I later published the finding as part of a larger study (Legéndy and Salcman, 1985), and the technique was tried out at other laboratories, the observation appeared to carry over to several regions in the brain, other than the visual cortex. As a diagnostic tool, the Poisson surprise is now used in some cases to detect diminished functionality, in others simply to furnish an additional functional parameter whose behavior can be linked to experimental conditions.

3.4 Critique of Brain Models Relying on *Average Spike Rates*

There is a large group of theoretical papers, including some well-known and important ones, which, because of their formulation, are blind to *surprising events* in neuronal spike trains, and therefore miss just about everything I describe in this book. These are the papers formulating neuronal inputs and outputs in terms of *average spike rates.*

In them, the *Computational Methods* section contains some statement along the lines that "the input to a (typical) neuron is a sum of the form $\Sigma\ W_i\ r_i(t)$, where W_i is the synaptic weight of the *i*th synapse on the neuron and $r_i(t)$ the average spike rate (as a function of time) arriving to the *i*th synapse."

Describing neurons in terms of well-behaved multi-component functions of time, like $r_i(t)$, opens up many powerful possibilities of mathematical and computer analysis (I also used *average spike rates* in the past, in a paper on columns and "subtractive normalization"; Legéndy, 1978); there is nothing wrong in using them when the objective is to make broad statistical statements involving ensemble averages. In addition, the existence of average spike rates $r_i(t)$ is not incompatible with correlations between the components $r_i(t)$, as long as they are slow, as implied in Fig. 2.3(b).

But if it is desired to include the effect of *surprising events* like the ones shown in Fig. 2.4, the "well-behaved" time dependence of the multi-component function $r_i(t)$ must be sacrificed, and with it most of the computational advantages of the formulation; in fact, to my knowledge, short-time correlation is never assumed in the average spike rates papers.

In other words, the practical (though unintended) effect of the *average spike rates* formulation is to slam the door shut on *surprising multi-neuronal events* altogether. However, as will be seen, a strong case can be made that surprising multi-neuronal events are the principal means of message transmission in the brain.

3.5 The LTP Is Probably the *Marking* of Synapse Sets for Later Use

In an addendum to the remarks on synapse modification in Sect. 2.4, let me weigh in against one more concept often seen in the literature (this time the experimental literature): the interpretation of long-term potentiation (LTP) as representing the neuronal version of long-term learning.

The observation casting doubt on the interpretation of LTP as the form taken by long-term memories, for instance memories lasting many years, has to do with the nature of spike-timing-dependent plasticity (STDP). Events of LTP and events of long-term depression (LTD) are believed to occur in roughly balanced numbers of synapses (Song *et al.*, 2000), and accordingly it is expected that the synapses potentiated in one series of events will sooner or later be depressed again in another, and in fact overwritten many times over the years by unrelated events of LTD and LTP.

In addition, when the LTP is described as a means of creating long-lasting responsiveness to a new combination of synapses (as in Fig. 2.4), the question arises: why is it that the LTP, according to all available data, only manages to increase EPSP amplitudes by about 60%? If the purpose of the LTP is to promote a set of synapses to a strength where they fire the neuron with certainty, one is well justified in asking, why isn't the synaptic strength increased 5-fold, or even 20-fold (Markram *et al.*, 1997)?

In view of such considerations, an alternative interpretation of the LTP arises: It is not intended as a means of strengthening a set of synapses and making the neuron responsive to them thereafter, but as a means of *marking* a set of synapses which bring reproducible input, prior to using them in subsequent processing.

From the standpoint of brain modeling, it can be stated that, in all the years it has been tried, no model of (non-trivial) self-organization ever succeeded where neuron-level learning consisted of the strengthening of synapses and nothing else. When all causes and effects were required to reside inside the neurons (as they are in the present model), the neurons needed to undergo more changes than mere synapse modification.

Among the sequences that are found necessary when one writes out the details (Parts 2, 3, 4), a typical one is where input arriving to one set of synapses prepares a cell for response to input subsequently arriving to another set of synapses (Sect. 12.5). The input from the first source does not require a response; the input from the second does; but the synapses of both sources must first be *marked* through repetition, to set their spikes apart from the noise. In general, it is a recurring theme in all the multi-stage processes below that when any of the component events shows itself to be able to rise above the noise, neurons must first *mark* the synapses on which it enters.

Such a concept implies the prediction that future research will reveal the "real" synapse promotion step, which is used only where the neuron is functionally required to respond to the synapses for a while thereafter, and which probably occurs in response to a second and independent set of molecules acting on synapses which recently underwent LTP, *dramatically* increasing the height of their unitary EPSPs.

The possible reason why the dramatic increase and other forms of synapse change do not show up without experiments specifically designed to seek them out is that the synapse-marking step is the first step of every sequence. It is only meant to separate the inputs from the noise and is presumably the step least demanding of specific preliminary processing. Experiments to date simply approach the cells with brute-force assaults on their synapses. While they do present a signal rising above the noise, they are, without specific preparation, only expected to accomplish the step designating the synapses as bringing signal rather than noise. The molecules which give rise to the next steps are only expected to arise in response to specifically targeted experiments.

If the LTP only *marks* the synapse sets, as I will assume in the rest of the book, it follows that the "Hebbian firing," by which I mean the action potential usually generated by a neuron undergoing LTP, while useful in the internal spreading of synaptic change within the neuron, is usually peripheral to the information sent out in the subsequent steps. (At most, it tells the other neurons that the neuron has "just noticed something.") In the rest of this monograph, therefore, the *Hebbian firing* is largely ignored.

The *Hebbian firing* is expected to be distinguishable from the information-transmitting firing in being less sharply synchronized, because it does not benefit from the volley-sharpening effect of *ignition* (Sect. 4.1), and is somewhat more tentative, because in its first few repetitions the dendritic spikes may not reach the axon hillock and generate action potentials.

Chapter 4
Cell Assemblies

Next we address the question of achieving synchronization without the benefit of a "system clock."

The answer in essence is that every task arranges its own timing. The signals relevant to a task only need to be heard by the neurons involved in the task, and only need to be sharply enough concentrated in time to be useful as trigger signals for the neurons that can hear them.

In other words, *the whole brain does not need to be synchronized.* It is enough to synchronize the spikes contributing to individual cooperative sequences.

Since a single neuron alone cannot be expected to synapse on all neurons that must participate in a task, the only way all those required neurons can be synchronized is by an arrangement where a whole set of neurons gives off spikes together, and the set is large enough that its members, between them, make sufficient contact with all the neurons which must be synchronized.

4.1 Ignition

The best known way for a set of neurons to give off coordinated spikes is through the phenomenon called *ignition*, which is a chain reaction-like spread of firing through a quasi-randomly interconnected group of neurons.

The term *ignition* is due to Rapoport (1952), who showed that a set of randomly interconnected idealized threshold neurons can, together, form a bi-stable device. If very few neurons fire, nothing interesting will happen; if enough of the neurons are made to fire (by some outside cause), the probability is high that the rest will catch on, and the whole network will continue to fire.

The ignition idea was combined with Hebb's (1949) "cell assembly" concept by Legéndy (1967), in a paper which introduced the crucial detail that a set of neurons (not necessarily neighboring ones) can be converted into an ignitable group by means of sufficiently strong connections between group members; once such a group is formed its ignition serves as its signature, able to broadcast its unique message to the rest of the network.

C.R. Legéndy, *Circuits in the Brain*, DOI 10.1007/978-0-387-88849-1_4,

The ignition idea, in its application to cell assemblies, remained virtually unknown until it was popularized by Braitenberg (1978), Palm (1982), Wickelgren (1999), and Scott (2002). As a point of terminology, it may be added that the papers of Legéndy (1967, 1970) called the ignitable group "compactum," rather than "cell assembly."

4.2 Synchronized "One-Spike" Ignitions

The version of ignition which appears in much of the literature (Legéndy, 1967; Palm, 1982; Wickelgren, 1999) continues to fire for a while after being ignited; this means that it does not pinpoint a single moment in time and is accordingly unsuitable for synchronization.

However, a minute change in the parameters (Legéndy, 1970) will modify the ignition mechanism and change it into a mode tailor-made for synchronization, because it lets every neuron give off one and only one spike per ignition, which means that all the spikes are emitted nearly simultaneously, and the group sends out volleys like the ones in Fig. 2.4.

The secret to the one-spike ignition (Fig. 4.1) is the "refractory period" after the action potential, the brief time interval during which the membrane is depressed and cannot support action potentials. To achieve one-spike ignition, the refractory period must be long enough to prevent the chain of excitation from making its way around the group and firing the cells a second time.

The synchronization achievable in one-spike ignitions is in principle much sharper than would be provided by the classical resistive-capacitive dendrites, thanks to a "temporal focusing" effect, operating via dendritic spikes, whereby the neuronal output is generated very fast after sufficient input has been received at the dendrites (Ariav et al., 2003); this means that when the input volley becomes sharply synchronized the output can be sharply synchronized to the volley.

In this writing, "ignitions" will always mean one-spike ignitions, where the cell assembly briefly falls silent after each of its neurons emits a spike. When a cell assembly is confined to a small (for instance 1 mm) piece of cortex, the spikes emitted in its ignitions are (typically) all within roughly 1–3 ms of one another; when the cells are farther apart, the time spread is wider, for instance 4–8 ms. In other words the ignition, by and large, gives off "near-simultaneous" spikes.

Since the summation interval of cells (roughly the relaxation time of EPSPs) is typically several dozen ms, which is longer than the typical time spread of the volleys caused by ignitions, every volley looks the same to the cells receiving them, regardless of the choice of cells which started it. Like the neuron itself, the cell assembly is a threshold device, and when its threshold is reached it does not matter how it is reached. The ignition conceals the history of its generation.

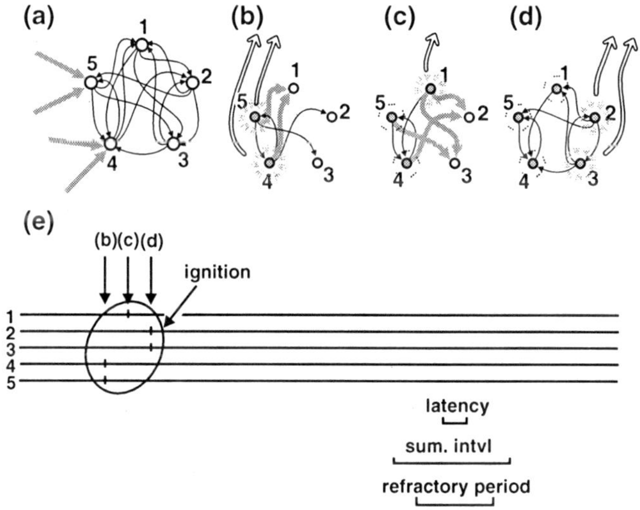

Fig. 4.1 Ignition of a cell assembly. Five neurons (1, 2…5) are interconnected (*arrows*), so that each neuron sends contacts to some (but not all) of the others, without regularity. The firing threshold in this illustration is assumed to be two, meaning that a neuron fires, after a short time (the *latency*), if two or more others send it spikes within a short enough time interval (the summation interval, *sum intvl*), and if it has not previously fired for a time equal to or greater than the *refractory period*. In (**a**), volleys are received by cells 4 and 5 from outside the group, sufficient to fire 4 and 5; in (**b**) 4 and 5 fire, and their output sends enough input to 1 to make it fire; in (**c**) 1 fires, and its firing together with the recent firing of 4 and 5 (neurons filled), sends enough input to the rest of the neurons, 2 and 3, to fire them; in (**d**) they fire. The recent firing of all the neurons sends enough input to neurons 1, 4, 5, to fire them again, but they do not fire because the refractory period after their last spike did not yet expire. In (**e**), the five spike trains are summarized showing the spikes as emitted in sequence

4.3 Ignitions and the Central Bins of Cross-Correlograms

It is well known that in cross-correlograms (Perkel et al., 1967), like the one redrawn in Fig. 4.2, the central bins which in this case correspond to time differences between −1 ms and +1 ms, sometimes contain a few extra counts (in the figure the number of extra counts is unusually large). The extra counts are used in experimental studies as evidence of functional linkage between the neurons.

The concept of one-spike ignitions places a new interpretation on the extra counts, namely that both of the recorded cells are members of the same cell assembly which ignites a number of times during the data collection. The spikes in the off-center bins can be interpreted as indicating that ignitions are not always ideally sharp; when the igniting cell groups have members located far apart in the cortex, the spikes are often spread out over 5–10 ms or more (Ts’o et al., 1986).

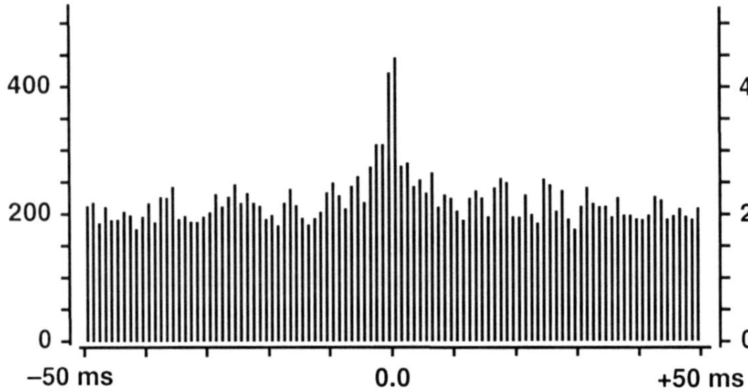

Fig. 4.2 The appearance of ignitions in a cross- correlogram. Correlation between two neurons picked up by the same electrode, from an awake cat. Associative cortex: *middle* suprasylvian gyrus. Histogram obtained over a 60 s period by gathering the counts into 1 ms bins, the height of the column at time t is the number of times the first neuron emitted a spike a time t ±0.5 ms after the second neuron (after Noda and Adey, 1970)

4.4 Ignitions and Single-Unit Recording

It will be noted that extracellular single-unit records do not reveal an ignition as a noticeable event because single-unit recording only looks at one cell at a time, and a person recording from the neuron within a cell assembly during ignition will only see a single spike.

In general, experimental verification of cell assembly dynamics requires simultaneous recording of multiple single neurons, and even when that is available it promises to present a challenge, since only a fraction of a local cell population is expected to participate in any ignition.

It may be noted, though, that the relative ease of finding cross-correlations between neurons close together (for instance pairs of cells picked up by the same electrode) indicates that the ignitions involve many more than two cells. If only the two monitored neurons showed correlation, detecting the correlation would require picking out exactly the right pair of cells, and that would seldom happen.

4.5 Why Myelin Is Indispensable to Nervous Function?

It is clear from the surprise-generating role of ignitions that these events are only useful if they can send a sharp volley of near-simultaneous spikes to all the cells they reach. If the time distribution of the spikes is broadened as a result of slow transmission, the ignition can become useless, because the neurons receiving it cannot distinguish meaningful volleys arriving to them from the background.

The requirement of simultaneous spike delivery becomes especially challenging when the ignitions involve neurons widely distributed over the cortical sheet. In the chapters below, visual integration of retinal images will be described in terms of ignitions which link groups often situated several millimeters (or centimeters) apart in cortex. The existence of correlations over such long horizontal distances in the visual cortex has first been reported by Ts'o et al. (1986), supplying the first indication of co-ignitions between neuron groups far apart in the cortex. (Once again, the finding indicates group-to-group correlation, because if only one neuron at each end participated, detection of the correlation would require picking out exactly the right cell at each end.)

Ignitions of widely distributed neuron groups present the problem that the differences between conduction delays from the various parts of the ignition will broaden the volleys delivered to neurons – unless spike propagation is very fast.

This is where the myelin, which speeds up spike transmission by an order of magnitude, makes its contribution.

Viewed in this light, the myelin, often regarded as being no more than an efficiency-enhancing device, shows itself to be an essential element of information processing. If the importance of well-synchronized volleys is as great as envisioned in this book, the myelin can make the difference between signals getting through and not getting through.

This in turn explains why demyelination diseases devastate the nervous system to the extent they do.

Chapter 5
Surprise, Statistical Inference, and Conceptual Notes

Let me say a few more words about the "surprise" mentioned in connection with ignitions, and look at the concept from another perspective.

Input to a cortical neuron arrives through axons whose originating cells in general are not individually listed in the genetic blueprint, except by their general cell grouping (cell pool). It is a good approximation that initially, in very early infancy, much of the incoming spike train appears to the neuron as gibberish.

In order to develop a useful response to the input stream, the neuron needs to have a way to detect statistical features of the spike train, because statistics are the only way to approach data not totally understood. The statement that neurons look for surprising events is equivalent to the statement that they seek to detect statistically significant properties of their inputs. The difference between the statements is mainly in connotation, as *statistics* are often associated with data gathered over days or years, and *surprise* with firing received inside seconds.

The reasoning which underlies the use of surprise is simple. The surprise imparted by an event is always computed on the basis of some probability distribution, and in some cases the surprise can be so high (the probability of the event so low) as to make the event amount to a miracle. In such a case (and in practice even in cases less extreme), it is justifiable to conclude that the assumed probability distribution is wrong.

When this happens, properly designed tests can often pinpoint certain compound events which, following the test results, must be assigned a probability much higher than they had in the original (baseline) probability distribution. In the brain theory context one is not interested in the updated values of the probabilities, only in the composition of the correctly chosen compound events.

When applied to statistics-based research, the "surprising event" serving as basis for the syllogism would encompass an entire experimental series from beginning to end, and the probability underlying the surprise would be the probability that the whole experimental series was a fluke, its results all accidental.

In the context of the visual cortex, the surprising events have short duration; they generally start and finish within a single "fixation" of the eyes. In that way they can be evaluated and utilized quickly. The determination of the compound event which is to be assigned a higher probability may, as in the example shown in Fig. 2.4, amount to marking the set of synapses on which coincident firing tends to arrive.

C.R. Legéndy, *Circuits in the Brain*, DOI 10.1007/978-0-387-88849-1_5,

5.1 Spike Coincidence Interpreted in Terms of Surprise

The ignition of a cell assembly sends out volleys which are surprising to the cells receiving them, if they receive spikes from enough members of the assembly (Abeles, 1982; von der Malsburg, 1999). That the volley sent out by an ignition is surprising can be easily seen by computing the probability that a number of *independently* firing cells send spikes inside a time interval which is for instance 1/10 the length of the average inter-spike interval.

It is not necessary to perform the computation to see that, with the parameters suitably chosen, single volleys can rise above the noise enough to justify short-term tentative recording, as implied in Fig. 2.4, and recurrence of the same (or nearly the same) volley a few times in a few seconds can confirm the event to a very high level of certainty.

It may be said, as a semantic point, that if a series of volleys like the ones in Fig. 2.4 repeats another dozen times, the repetitions will "stop being surprising after a while." For, it is then "clear" that the volley is not as infrequent as originally expected. In other words, saying that the surprise of the compound event should keep increasing by undiminished amounts with each new repetition goes against the common-parlance usage of the word "surprise." This is in fact quite true, and in this respect the brain-related usage of the word diverges from the everyday usage.

Simple repetition of volleys of near-simultaneous spikes is one example of the surprising events discussed below and another one is large overlap between successive volleys, as occurs when the set of co-active neurons gradually shifts, because the retinal image drifts (Fig. 13.1(b)). More complex examples include multistage setup protocols (Sect. 12.4) where volleys coming from a series of interrelated neuron pools in quick succession turn on the marking of synapses for inputs expected from yet another neuron pool.

Each time, when signals are to cause change, they must rise above the noise through surprise inherent in them, and the pieces of tissue changing must be able to peel the signals away from the noise by biological procedures which amount to detecting the surprise.

5.2 *Local Knowledge* and Its Relation to Information

Brain function revolves around information, and it is essential to be able to keep track of information flow between neurons and groups of neurons. As it happens, Shannon's (1948) information measure does not offer the best way to describe information flow in the brain, because that measure overestimates and somewhat mischaracterizes the information content of signals between the cells.

Besides, in brain theory it is useful to be able to identify the *nature* of information conveyed by surprising events, and spell out what each event "tells" the neurons it reaches (as in Lettvin's famous title).

The term "knowledge" or "local knowledge" has advantages in dealing with the brain, because *knowing* is a transitive verb; it naturally reserves a place in the sentence for the thing *known*. After the signals "tell" a neuron something, the neuron is in a position to "know" it. (The neuron does not automatically know it; it only does if it is also *equipped to detect* the information and detects it.)

The assumption underlying the verbal notation is that, provided the needed information can be gotten to the neuronal localities, the neurons, thanks to chemistry and genetics, can manage to change or act *as if* they knew exactly what they were doing. Accordingly we, the persons studying the system, can ascribe "implied knowledge" to every neuron, and speak of it as if it were a "*scient object.*" After all, both the causes and the effects of the *implied knowledge* are parallel to those of literal knowledge.

As a separate matter, it may be noted that "knowing," as the term is used in common parlance, carries a connotation of certainty, or a probability equal to 1.0. In brain theory the level of confidence is less. However, as discussed below (Chapter 7), the network tends to provide a way to keep improving the confidence until the probability comes very close to 1.0.

5.3 The Fundamental Law of Brain Theory

Of the many pieces of data and logic which can shape a brain model, arguably none goes further toward defining a concrete network than two partly vague and partly obvious statements: *When neurons fire or undergo change, it is because they know that they have to, and no neuron can possess any knowledge that has no way of getting to it.* The first of these relies on the implicit assumption that the brain, for the bulk of its work, does not rely on trial and error (because it would take too long).

One is tempted to call the twin statements the "fundamental law of brain theory," because of the remarkable degree to which they can narrow down the possibilities. Alternatively, one can just say that any realistic model of brain function must first and foremost be "geometrically possible," in the sense of providing the pathways and methods to support the "logistics" of operating information. Samples of thought processes driven by the twin statements and their logistic consequences are presented in Sects. 19.1 and 20.1.

It may be mentioned, in this connection, that Hebb's (1949) classic synapse modification rule is geometrically possible. The proposition that correlation between presynaptic and postsynaptic activity should tend to reinforce a synapse does not require any information to get to the synapse that cannot get to it.

5.4 *Parsing* the Network into *Localities*

The *local knowledge* concept can be used for breaking down the system into component parts for easier handling, because not only the neurons, but all pieces of brain tissue, large and small, need to know what to do before doing it.

The device of a "black box" is widely used in electrical engineering for breaking complex systems into smaller parts, and a variant of it is useful in brain theory. It envisions a locality (for instance set of neurons) as being surrounded by an imaginary wall, to form a "black box," and makes the optimistic assumption that, if the apparatus inside the black box has the necessary information at its disposal, in the form of incoming surprising events interpreted with the help of chemistry and genetics, then the apparatus in the black box, through the action of the elements in it, will manage to do its part in the overall processing.

Replacing neuronal localities by black boxes, and concentrating on what each box can know based on its inputs, offers a way to "parse" the neuronal network. It *cuts off the inquiry* at the level of the boxes and allows us to ignore the problem of modeling the machinery inside them. By use of the black box concept, all pieces of nerve tissue, from small pieces of nerve membrane to whole cortical regions, can be treated as if they were *scient objects* without any further worry about their internal workings.

An example of breaking down the same problem to two different levels of detail is offered in Chapter 20, where Fig. 20.1 parses down the solution of a processing problem to the level of cortical columns, then Fig. 20.2 parses it down to the level of neurons and synapses.

Parsing, in the context of neuronal networks, is a matter of practical necessity, without it the subject becomes too large. In the chapters below, communication will be parsed down (mainly) to the level of single neurons or groups of neurons. While the parsing allows us to bypass the question of *how* the "black boxes" operate, it does leave us with requirements as to *what* they have to do, and allow us to judge whether the requirements are reasonable.

5.5 Brain Modeling Viewed as "Reverse Engineering"

The concepts of surprise and local knowledge deal with events and their causes as viewed from the perspective of the neuronal locality.

The other half of brain modeling is to look at the same situation from the opposite perspective, where we start from some fact of the world or the network, and, in a kind of *Gedanken Experiment*, design a system which can effectively convey the fact to given neuronal localities by means of surprising events.

In other words much of brain modeling is an effort of "reverse engineering," which in practice is best accomplished with an attitude of viewing things from the standpoint of the engineer. How to convey to the equivalence of these signals to those neurons? How to speed up the response of the network?

The results, once obtained, are of course easily rephrased without reference to engineering principles (and rendered more palatable to dogmatists), but I will not bother. The purpose of theory is to impart understanding, and to conceal the engineering beneath the processing can only obscure the causal linkage between structure and function.

Chapter 6
A New Term: Ignitions Which "Reach" or "Don't Reach" a Neuron

The word "reaching" will be redefined in the present context, in terms of the surprise imparted to a target structure.

It will be said that *an ignition (or a compound event consisting of several recurrences of an ignition) reaches a neuron* if it imparts enough surprise to the neuron to enable the *marking* of a set of synapses on it (Fig. 4.2). By extension of the same idea, *an ignitable group reaches a neuron* if its ignitions reach it, and *an ignitable group reaches a group of neurons* if it reaches the "average" neuron of the group.

Although these definitions deal with signals, *reaching* is, to a great degree, a property of the physical connectivity of the network. It is, by and large, an anatomical (morphological) concept. A neuron group reaches a neuron if it makes enough synapses on the neuron.

The "fundamental law of brain theory" can then be rephrased to state that whenever pieces of brain tissue act or change, they must first be *reached* by signals telling them to do so.

Reaching has relevance to neurology; for instance the "disconnexion syndromes" of Geschwind (1965), which are neurological deficits resulting from damage to fiber tracts between cortical regions, can be described in terms of failure of neuron groups in one region to reach the neurons in another.

As was noted in the Introduction, if one were to name the single most noticeable feature of the brain, one would probably choose the rich branching of its neurons, both at their input end and their output end, and the great wealth of processes between neurons, long and thin, with thousands of contacts. The "average neuron" has been estimated to have some 38,000 synapses in humans (Cragg, 1975), a huge number. In the mouse the corresponding number is about 8,000 (Schüz and Palm, 1989). In the cat (Binzegger *et al*, 2004), for layer 2–3 pyramidal cells (which are somewhat smaller than the "average neuron") the number is reported to be about 7,000.

It is clear from the extensive connectivity that considerations of *reaching* are central to brain design.

C.R. Legéndy, *Circuits in the Brain*, DOI 10.1007/978-0-387-88849-1_6,

6.1 How Many Synapses Does It Take to Reach a Neuron?

The point to note is that a single synapse is not enough for reaching a neuron. To be sure, repetitions of volleys can increase the imparted surprise, but the individual volleys must still carry enough surprise to rise above the noise and initiate the process of marking a set of synapses (Fig. 2.4); and after a synapse set is marked, subsequent input events must be substantial enough to fire the neuron. Reaching requires a large enough number of synapses.

The presently available evidence most relevant to the issue of reaching comes from studies of long-term potentiation (LTP). In the recent experiments of Remy and Spruston (2007), the LTP was induced by 5–8 mV excitatory postsynaptic potentials (EPSPs). By assuming that the amplitude of the typical unitary (single-synapse) EPSP was 0.2 mV, the authors estimated that in their experiments about 25–40 synapses had to be simultaneously active, all of them on the same dendrite, in order to create observable LTP. The more direct experiment of Losonczy and Magee (2006) puts the number at 20 (both studies are done on pyramidal cells of the hippocampus).

It is hoped that both figures will prove to be overestimates, and that input-starved neurons will be found able to change their thresholds temporarily, greatly reducing the required number of synapses. One possible mechanism is that the neuronal membrane dynamically regulates the required intensity of volleys, by tentatively raising the membrane potential (possibly by temporarily opening ion channels to permit positive ions into the cell interior), thereby effectively lowering the thresholds, until the available inputs occasionally cause dendritic spikes and firing. (A comparable assumption is described in connection with Fig. 7.1 below.)

When the resulting firing is caused by a volley able to repeat itself, the resulting short-term potentiation will be promoted, after some repetitions, and tentatively mark the synapses. Subsequently the thresholds can be raised again (the membrane potential lowered), and the selectivity of the process restored. In the cases where reduced selectivity causes potentiation to occur in error, the participating synapses remain potentiated in the short term, but soon return to their original state after the thresholds are restored, for lack of confirming input. (The spikes produced in this way contribute to the random component of neuronal firing.)

It must be added that even if the input requirement is greatly reduced in future estimates, the requirement will remain that "at least a few" members of any igniting group must contact a neuron if its ignitions are to reach the neuron.

As will be seen, for the purposes of the present qualitative study this much detail will suffice.

6.2 Axons, Where They Arborize, Can Probably Contact Most Neurons

One important number which can be estimated using the available anatomical data is the probability of synaptic *availability* between a neuron and an axon arborizing near its cell body.

An axon is "available" to a neuron, and vice versa, whenever the axon passes close enough to a dendrite of the neuron to be within reach of the dendritic spines. The term "potential synapses" (Stepanyants et al., 2008) is often used in the literature for my "available" synapses, but I will continue to use "available," because it fits conveniently into phrases like "the number of neurons available to average axon arborizing in a column."

Availability is not the same as synaptic connectedness; in fact it is known that not all axons available to a neuron actually make a synapse with it. But it will be assumed that when it is functionally desirable for mutually available processes to have synaptic contact, the contact will always exist.

Therefore in questions of reaching, where functional contact is desirable by definition, the number of contacting axons can be equated to the number of available axons.

As Fig. 6.1 shows, there is a fairly easy way to estimate the probability that a given dendritic tree is available to a given axon. One draws out the silhouette of the dendritic tree, widening the dendrites to extend to the farthest axon center within

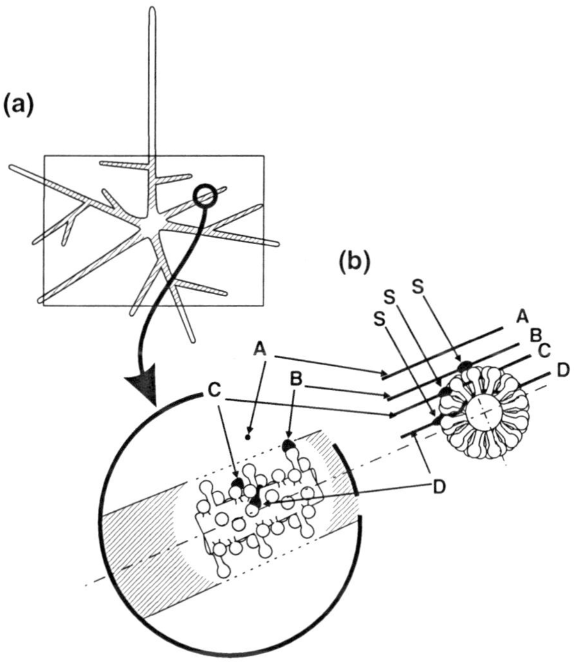

Fig. 6.1 Graphical estimation of availability ratio. In the inset, a magnified version of a segment of a dendrite is shown, in two views, to illustrate the extent of the *shaded area* in relation to the dendrite and its spines. Four axons are shown (blackened, schematically drawn to be arrow-straight): A misses the dendrite; B, C, and D contact it. An axial view of the situation is shown next to the inset to illustrate how the four axons run by, and how three of them make synapses (S) with the spines. Axon D is displaced by the dendritic shaft and goes around it. The straight portions of the axons are only visible in the axial view, and are only seen as small dots in the main view (see A), where they run at right angles to the plane of the figure

reach of the dendritic spines, and then planimetrically estimates the fraction of area covered by the silhouette. It is easy to see that the fraction is equal to the probability that a fictional arrow-straight axon, perpendicular to the plane of the drawing and traversing through the dendritic tree once, is available to the neuron.

It will be noted that the direction in which the axon runs does not make much difference (anisotropic connectivity does exist in the cerebellum, but is probably not important in the cortex). Therefore the estimation reduces to determining the amount of axonal length (including all branches) traversing through the dendritic tree and expressing it as an equivalent number of times the axon traverses through the dendritic tree.

If the silhouette occupies 15% of the dendritic area as in Fig. 6.1 (an overestimate), and the axon with its branches effectively traverses the dendritic tree 5 times (often an underestimate), the probability that one traversal misses is 1–0.15, the probability that all 5 traversals miss the dendrites is $(1–0.15)^5 = 0.44$, and the probability that there is at least one potential synapse is 1–0.44, or 56%.

Since 5–10-fold traversal of dendritic trees is more or less typical in axons where they arborize, it can be said, by and large, that *axons can (if they need to) make synapses on most of the neurons whose cell bodies are within their field of arborization*. The careful measurements of Stepanyants et al. (2008) lead to a similar conclusion.

6.3 A Good Unit of Cortical Distance: The Width of a Column

The width of the dendritic trees in cortical pyramidal cells is of the same order of magnitude as the extent of the local branching of their axons, and is of the same order as the typical width of cortical columns, roughly half a millimeter. It may be asked whether this distance has any functional significance.

A piece of insight into the significance of this distance is offered by the results of cortical microstimulation experiments (Bak et al., 1990; Schmidt et al., 1996). In these, as part of an effort to create visual prostheses for the blind, the human visual cortex has been stimulated with small electrodes, sending electric currents into the immediate neighborhood of the electrodes. Such microstimulation results in the visual impression of small sometimes colorful dots, or sparks, or bars of light ("phosphenes").

One of the findings was that when two stimulating electrodes were used their phosphenes could (usually) only be recognized as being separate when the horizontal distance between the electrodes was greater than about 0.5 mm; if they were closer together they (by and large) appeared as single phosphenes. The significant detail was that this distance was apparently independent of the distance of the electrode pairs from the central (foveal) projection; in other words it was independent of the retino-cortical magnification factor!

Since 0.5 mm is, as mentioned, roughly the width of local neuronal arborages, and also of the cortical columns (hypercolumns, by the terminology of Hubel and

Wiesel, 1968), or alternatively the distance between the centers of the cytochrome oxidase (CO) blobs (Livingstone and Hubel, 1982), the finding ties up a number of loose ends in our understanding of vision.

It tells us that the cortical column, or the CO blob, is in essence a "point" in the visual cortex, and the average distance between the centers of columns represents a crude measure of visual resolution. It may be called the "cortical distance unit."

The statement identifying the column width with visual resolution is very crude, and is only legitimate for the purposes of a simple model like the present one. In fact the visual acuity of a person with good vision is somewhat finer than the column size, and the vernier acuity, at which a fault in a straight line can be recognized (Westheimer, 1976) is finer yet.

It is likely that the higher acuity achieved in actual vision utilizes data collected over a number of columns and refined through averaging, but data refinement through averaging is beyond the scope of this writing.

6.4 Axonal Branching Near the Cell Body

One known feature of cortical neurons is that their axons usually make a number of branches around the cell body ("recurrent collaterals"). As was seen in Fig. 6.1, such a pattern of arborization renders most of the nearby neurons (within about 0.5 mm, as mentioned) *available* to the axon.

Since one of the requirements of ignitability is that every member of the ignitable group be reached by the sum-total of axonal processes from the rest of the group, the recurrent collaterals make the formation of ignitable groups within small cortical neighborhoods (local ignitable groups) feasible. Local ignitable groups make especially efficient use of the brain's hardware resources, as the total axonal footage required to create an ignitable group is smallest when the members of the group are close together.

The postulated existence of local ignitable groups is consistent with the finding that cross-correlograms showing extra counts in the central bins are often obtained from pairs of neurons picked up by the same electrode (and separated on the basis of spike amplitude). In other words, from pairs of neurons close together.

6.5 Retinotopic Mapping

Some known features of cortical neuroanatomy can be looked upon as the brain's way of helping neurons *reach* their target neurons, while using as little fiber footage as possible. One notable example is supplied by the phenomenon of retinotopic (and in general somatotopic) mapping.

The principle is illustrated in Fig. 6.2, where two neighboring neurons are envisioned to send axons to one part of the brain to another. If each axon were to distribute its axonal branching widely in a cortical area, the likelihood of both axons

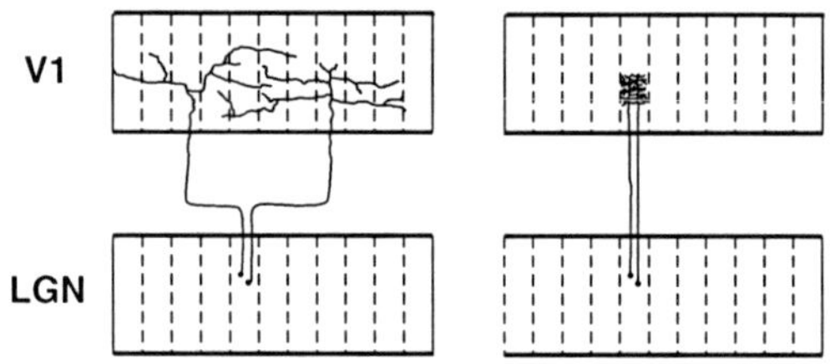

Fig. 6.2 The trick of retinotopic mapping. The drawing shows two neurons envisioned as sending axons (for instance) from the lateral geniculate nucleus (LGN) to V1. If both neurons were to spread their axonal arborages widely across the cortical sheet (*left-hand drawing*), almost no neuron would be within contacting range of both axons, but if they spread them over the same layer of the same column (*right-hand drawing*), the majority of neurons there will be available to both axons. These drawings, and other similar "anatomical diagrams" below, are hand drawings intended for illustration only and not to be considered as data

synapsing on any neuron would be small. By concentrating the contacts, the neurons can ensure that, at least in the column somatotopically corresponding to a local ignitable group, many neurons will be contacted by both axons. It is noted that in addition real axons tend to arborize in a limited set of cortical layers, which concentrates the fibers even further.

It is likely that without the trick of mapping, the messages generated by local ignitions in visual area V1 would remain stuck there; they would never be able to make their way to V2.

6.6 Axons Which Confine Their Branching to a Few Columns

Another noticeable feature of cortical neuroanatomy, which can be interpreted along similar lines, is that many long axons often do not make branches throughout their length, even less so throughout the whole cortical area, but concentrate their axonal arborage over one or a few regions, each comparable in size to a cortical column, and usually within one cortical layer.

Figure 6.3 shows the advantage of the concentrated fibers in terms of reaching. When a collection of neurons forms a local ignitable group, and each member of the group sends its axon far and wide through the cortical area, hardly any neuron will be reached by the group because hardly any will receive synapses from more than one member. When the same group of neurons arborizes in a few distinct regions, ones functionally desirable for the group to reach, the likelihood that they succeed can become significant.

In summary, one may say that the axons obey two simple rules which, combined, justify the statement that *wherever neurons need to be reached they will, by and large, be reached*. The rules may be called the *concentrated arborization rule*, which states that an axon arborizing in a neighborhood will make enough branches in that neighborhood to *contact* the majority of functionally compatible neurons there, and

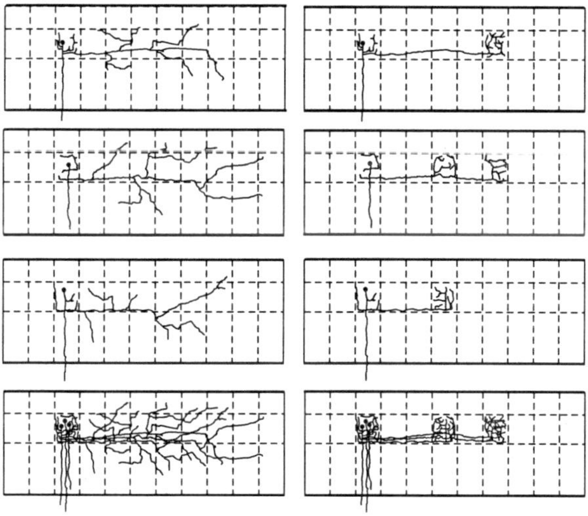

Fig. 6.3 The trick of concentrated arborization and convergence. The cortical sheet is represented schematically as a parallel slab in which column boundaries are marked out for sake of reference. If members of a local ignitable group all arborize widely throughout the cortex (*left-hand drawings*) they will reach few if any neurons, but if they arborize in a few selected columns (*right-hand drawings*), they will tend to reach the neurons in them. The drawings at the *bottom* are superpositions of the drawings above them

the *convergence rule*, which states that when an axon arborizes in a neighborhood, it will not be alone in doing so, because enough other neurons will concentrate their axonal branches there to *reach* the majority of neurons it is functionally desirable for them to reach.

Chapter 7
Confirmation Loops, Powered by Self-Ignitions

7.1 The Principle of Overwhelming Odds

One of the central edicts of brain theory may be called the "principle of overwhelming odds." It states that whenever a message is important, the system has a way to package it in events so surprising that it is "essentially impossible" for the events to be accidental. For instance the expected frequency of their accidental occurrences is once in many days or years.

In this connection, it is necessary to emphasize (Sect. 2.4) that an event of surprising input to a neuron does not necessarily act to *fire* the neuron receiving it. It may simply mark a set of synapses whose role must await a later input received by another set of synapses, or alternatively it may place the neuron into a new state. All these inputs to a neuron, many of them in the nature of "control" inputs, must *reach* the neuron, and in each case this means that the event conveying them to the neuron must be highly surprising.

The important thing to realize is that a single volley from an igniting cell assembly usually does not impart enough surprise to satisfy the principle of overwhelming odds. It follows that the only way an ignition can be made into a sufficiently "forceful" message is through several repetitions; the ignition must keep repeating itself until its message is effectively "hammered in" (Fig. 2.4).

The need for repetitions may be obvious from the standpoint of the neurons which *receive* the signals; what is not immediately clear is how the repetitions are implemented by the circuits that *send out* the signals. The point being that every one of the repeatedly received volleys must come from some network able to *generate* it.

7.2 Prime Mover Networks at the "Sending End" of Surprising Signals

The realization that some neuron groups must be capable of sending out repeated volleys introduces a new concept: The brain must contain networks that can repeatedly go into instability when called upon to do so. Further, such networks must be built into every stage of the information processing system; in other words not only the motor division, where the need for them is to be expected, but the sensory division as well.

C.R. Legéndy, *Circuits in the Brain*, DOI 10.1007/978-0-387-88849-1_7,

The repetitions generated by the unstable networks will be referred to as "confirmation loops," because the purpose of each repeated volley is to confirm the message of the previous ones.

Since all the surprising firing events must be manufactured inside the brain, each confirmation loop must have a "prime mover," in the form of a cell assembly that can be made to keep igniting on its own, until its ignitions are no longer needed.

The repeated instability in such a cell assembly will be referred to as "self-ignition." Its mechanism is illustrated in Fig. 7.1.

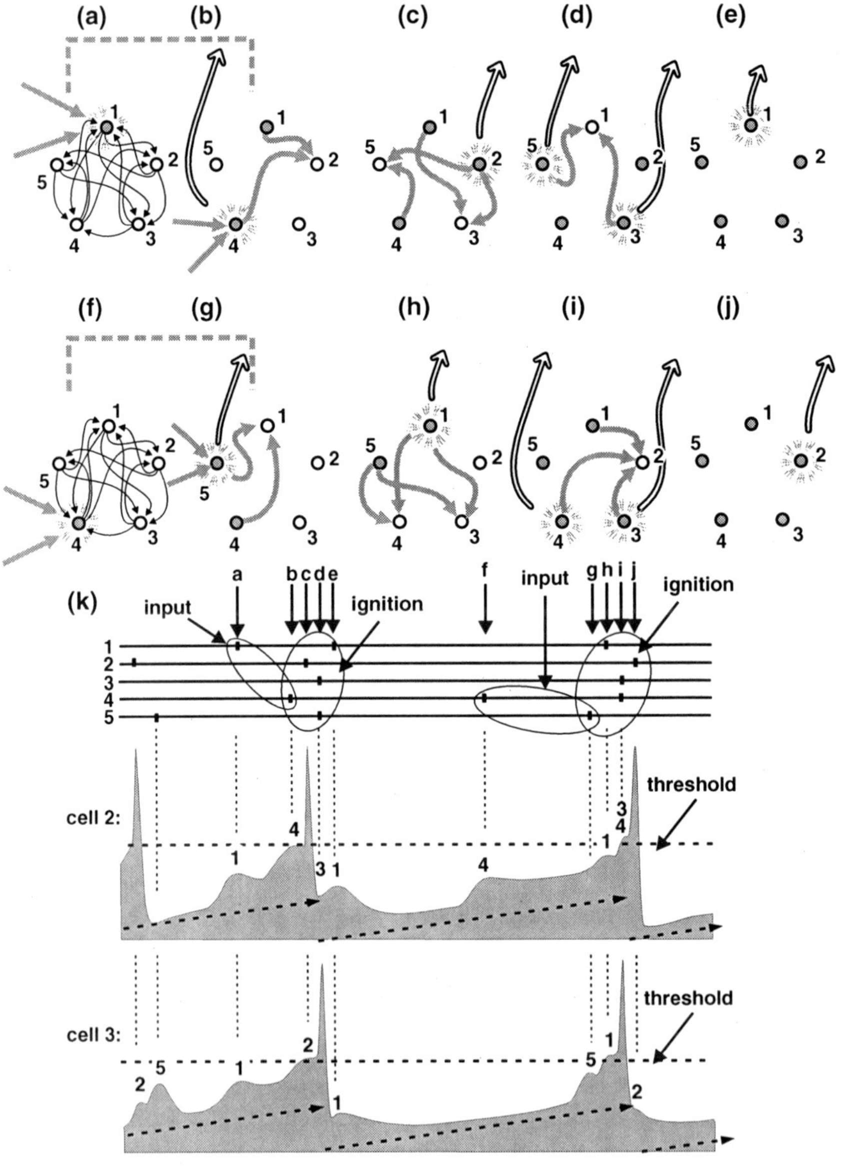

Fig. 7.1 (continued)

It will be noted that each individual neuron in a self-igniting assembly contributes a single spike to each repetition, no more, and that the ignitions do not need to occur in very quick succession; they only need to occur at a rate much greater than the accidental rate of comparable volleys. Accordingly, confirmation loops do not cause a dramatic increase in single-unit spike rates.

The challenge in designing an acceptable mechanism of repeated self-ignitions is that it is undesirable for the self-ignitions to occur periodically. They should occur on a random schedule.

The random timing is imposed on the model by observations of cortical firing. If repetitions occurred periodically, the typical member of a self-igniting group would fire periodically, and periodic firing would often be seen in the cortex – which it isn't (Sect. 7.3). In addition, the random timing is an essential element in the elimination of crosstalk between distinct contours (Fig. 19.6), and in echo detection within a noisy environment in linkup algorithms (Fig. 20.2).

The ignitions are initiated by the random firing of the cells (arrows from left, converging on cells in (a), (b), (f), (g)), and the assumed gradual rise of intracellular potential. In the illustration both ignitions occur as a result of two spikes arriving in relatively quick succession (circled as "input"); however, in the absence of such coincidences, according to the assumption, after a little while the rising membrane potential would also cause single spikes, or mere noise, to initiate ignitions.

7.3 Confirmation Loops and the Classical "Reverberations"

Confirmation loops are similar in their effect to what Lashley (1930), Lorente de Nó (1938), and Hebb (1949) described under the heading of "reverberations," the postulated tendency of neuronal excitation to go round-and-round among interconnected cells, thereby causing the cells to fire repeatedly.

The synapse modification in Hebb's scheme was envisioned along the lines of simple conditioning, and the reverberations were used as a means to provide the repetitions which would support a microscopic version of such conditioning (one-trial learning was contrary to ruling dogma at the time). Arguably, the rationale underlying the repetitions translates to the same rationale as used in this writing, the need to compound the surprise arriving to localities.

Fig. 7.1 (continued) Generation of randomly repeated self-ignitions. The figure illustrates the way in which slow charge leakage into the cells of a cell assembly can lead to repeated instability and ignition of the assembly. The simple cell assembly used is the same as in Fig. 4.1. Two ignitions are shown in the sequences **(a)**–**(e)** and **(f)**–**(j)**. The spike trains, including the two ignitions, are plotted in **(k)**, with letters a, b,. . ., j at the top referring to the drawings where the spikes are emitted. Underneath, the hypothetical time course of the membrane potential is sketched for two of the cells, 2 and 3. For illustration, the sketches show the membrane polarization as if it were positive. The numbers above the unitary EPSPs refer to the cells whose firing caused them (noting that the cells only receive input from a subset of the other cells, as shown in (a) and (f)). Slanted upward-pointing arrows show the gradually rising baseline potentials caused by the charge leakage

The contribution of the present writing to the decades-old reverberation discussion is in introducing the one-spike ignition and combining it with the idea of randomly timed self-ignitions.

As mentioned, periodic firing is notably absent in the cerebral cortex. In fact, the absence of periodic firing was frequently cited as an argument to rebut the concept of reverberations during the 1950 s.

I may mention that I had at some point devoted more than a year to an unsuccessful attempt to detect brief periodic spike sequences in the visual cortex with the help of rigorous computer-aided spike train analysis. In an unpublished part of a 1980 s study (Legéndy and Salcman, 1985), I had tried to find statistically significant epochs of periodic firing in single-unit records taken from the visual cortex of behaving cats, and failed. The negative finding was puzzling, since at the time I already knew about the role of surprise in the information traffic within the brain, and the probable need for repeated volleys, yet the results seemed to disprove presence of repetitions. Only much later did I realize that, by using a trick like the one illustrated in Fig. 7.1, it was possible to envision confirmation loops with *irregular* inter-volley intervals.

Chapter 8
Communicating "Relatedness" Through Time-Linked Ignitions

It is a safe guess that the old view of the visual cortex, which envisions it as sending no more than a structureless collection of edges and bars to higher cortex, misses something important. It is clear that, in some way, the information stream sent out from the primary cortex must tell the higher cortex how to synthesize those elements into shapes; in other words it must communicate to it, in some form, the *relations* between the simple figure elements; it must include what may be called the "figural syntax."

The challenge here is that the language of neurons is not very expressive. Neurons, for instance, cannot "point" at other neurons; all they can do, by and large, is choose when to fire.

However, as it happens, that is all they need to do. While one neuron cannot speak to another neuron about a third neuron, two neurons can form a link with which to synchronize their firing, and arrange among themselves to tell a third neuron that their firing is related.

In the terminology of von der Malsburg (1981, 1999) this is an example of *temporal binding*; in fact, as mentioned in the Introduction, the present book can be regarded as one addressing the *binding problem.*

In general, if neuron groups are assigned to verbally describable "objects" of the outside world and two of the objects are "connected" in a metaphorical sense, the brain tends to translate the metaphorical connection to a hardware link implemented through fibers and synapses.

Once there is a physical connection, it can help the groups ignite in a time-linked manner, and thereby announce the connectedness of the two real-world objects to all neurons they reach.

The simplest form of temporal linkage is simultaneity (which in practice is *approximate* simultaneity), and the main connection-conveying method explored in this model will be near-simultaneous ignition ("co-ignition") of two or more ignitable groups.

C.R. Legéndy, *Circuits in the Brain*, DOI 10.1007/978-0-387-88849-1_8,

8.1 Time-Linked Ignitions Viewed as Sentences

Viewed in isolation, time-linked ignitions between two neuron groups can only convey that their corresponding objects are "related." It is up to the molecular signaling pathways, inside and near the cells, combined with genetic clues, to communicate the nature of the relatedness, based on the biological identity of the neuron pools involved. The linked ignitions deliver visual sentences, and the different sentence parts are conveyed by ignitions occurring on different neuron pools, each identifiable by its specific physiological signature.

In some of the cases described below, the co-ignition tells the cells it reaches that the participating cell groups have their receptive fields *on the same contour*. When this is so, the circuits detecting the co-ignitions process them in a way consistent with having received same-contour information.

The occasional reference here to grammatical syntax is meant to be more than a play on words; it expresses an assumption that spoken language has developed from what is in fact the internal communication method of the brain. The time-linked ignitions are not similar to sentences, they *are* sentences.

Speech is limited by its nature to uttering the words sequentially, and accordingly the sentences are "packets" of words small enough that the relations between the words uttered in sequence are still easy to keep straight. In a simple protocol of sound and breath, brief pauses made between sentences and the characteristic intonation of speech keeps the sentences from blending into one another.

The present model uses the time-linking of ignitions into co-ignitions in the same way as speech uses the grouping of words in sentences; both methods package sets of elements into groups which are small enough to avoid ambiguity.

In the brain, many sequences of sentences can be transmitted at the same time; this is the brain's version of *parallel processing*. The use of distinct neuron pools is enough to avoid ambiguity inside sentences, but not enough to avoid interference between them.

Additional devices for keeping separate processes from interfering include sufficient cortical distance between comparable ignitions and sufficiently low occurrence rates, both of which reduce the probability of interference. In addition, through the irregular timing of confirmation loops (Fig. 7.1) the likelihood of confirming a parasitic linkage can be made small (Fig. 19.6).

8.2 Joining Sentences on Shared Nouns

Attempts to write a model of relations between objects invariably run up against the apparent contradiction between the large number of interrelated objects of interest and the small number of objects (often only two) connected by any one relation. Spoken language has a simple but powerful solution to the problem, and, as will be seen, the same solution can be applied to vision.

One remarkable achievement of spoken language is that it can join the message of many sentences into a single larger message. The joining is possible thanks to the

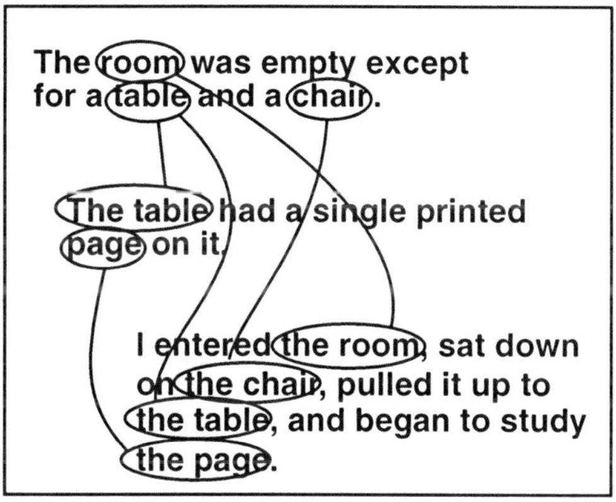

Fig. 8.1 An example of sentences joined on shared nouns

ability of languages to indicate when a noun used in one sentence denotes the same object as a noun used in another later sentence.

When, in English, we speak of "*the* table" rather than "*a* table," it generally means that we refer to the same table as we did in some previous sentence.

For instance, let us consider the following three sentences linked through nothing other than nouns in conjunction with the definite article (Fig. 8.1). "The room was empty except for a table and a chair. The table had a single printed page on it. I entered the room, sat down on the chair, pulled it up to the table, and began to study the page." The sentences are linked multiple times, on the nouns "room," "table," "chair," and "page." Nowhere is there any doubt that the three "tables," for instance, are one and the same object.

The definite article (which is absent from many languages) is one grammatical device for transferring noun identity between sentences; gender-matched pronouns are another; noun phrases and proper nouns, which can transfer noun identity between sentences separated by many pages, are others. They all answer the same need: to convey a picture with too many details to fit into a single sentence.

The trick of joining sentences on shared nouns is of great interest here because of the role we attribute to it in vision. Further, the idea, in all its trivial simplicity, is a crucial step in solving the *binding problem*, because it shows that the small number of elements tied together in elementary linkages does not set a built-in limit to the complexity of pictures the linkages can handle.

The visual field contains the images of many objects and each image can be made up of many details. The relations between these are visual sentences, transmitted in pieces, partly in sequence (in no particular order) and partly in parallel, and the pieces are joined into larger pictures on their shared nouns. At the receiving end in the brain the shared noun objects are recognized as being the same by a combination of tracking (Sect. 13.1) and the selection of cells announcing them.

An example of a noun is the *node* (Chapter 15), which represents, essentially, a selected point in a contour. Within nodes, certain cell groups also play noun roles. Whole contours and color patches are also nouns, but their treatment is not taken up in this model.

Chapter 9
Relational Firing: Broadcasting a Shape Through Time-Linked Ignitions

In the next few sections I will jump ahead of the story and attempt to illustrate how co-igniting neurons can convey shape information, provided they first manage to find each other and link up. (The linkup is discussed in Chapters 19 and 20).

9.1 Labeled Lines: The Messenger Is the Message

In the visual context, the sentences are simple and their form is repetitive. One sentence of interest below will be the co-ignition which transmits the message that two points bear a certain directional relation to one another, along the lines of "point 1 is in the direction of *ten o'clock* relative to point 2."

Here the "ten o'clock" part is in the nature of relative direction between points, in other words the direction in which one point is displaced relative to the other. The information contained in it is the sort contained in a *unit vector*, which is a vector conveying no magnitude information, because its Cartesian components are merely the directional cosines.

The principle through which an ignition can transmit relative direction may be considered as being an extension of Müller's (1834, 1840) old "principle of specific nerve energies," or its more current version, the *labeled lines* principle (Reich et al., 2001).

At present, *labeled lines* research is concentrated on distinctions between peripheral sensory nerves and somatosensory modalities (see for example Belmonte and Viana, 2008), but it may be hoped that in the coming years it will become possible to explore the version of labeled lines of interest to this book, where pools of (presumably) glutamatergic/aspartergic synapses, on axons from differing cortical neuron pools, are helped by molecular signals to exert effects specific to the neuron pool of their origin.

The most widely accepted example of the generalized version of labeled lines is the concept of conveying *orientation* information by the axons from the neurons having the corresponding orientation preference. Superficially, the axons carrying different orientations appear similar, yet they cannot pass on their orientation information, and their orientation information goes wasted, unless the cells they contact can recognize their synapses as representing their specific orientation.

C.R. Legéndy, *Circuits in the Brain*, DOI 10.1007/978-0-387-88849-1_9,

In this form, the *labeled lines* principle states that some neurons communicate their message through their recognizable identity; in other words recognition of the message is accomplished by recognizing the neuron pool whose axons bring it. *The messenger is the message*. In practice, the messenger is only a *part of* the message, and other parts are communicated through the synchrony with other neurons and in other ways, but the messenger does carry a significant part of the transmitted information.

The second example of the same principle, which provides the springboard to this model's discussion of shape processing, does not have the experimental backing of orientation coding; rather it is suggested by the likely availability of a mechanism morphologically ideal for its implementation (Figs. 9.1 and 9.2). Neurons of this example, the "direction-coded cells," communicate "relative direction," which is the orientation of an imaginary straight line drawn from the retinotopic location of the cell body to another retinal location.

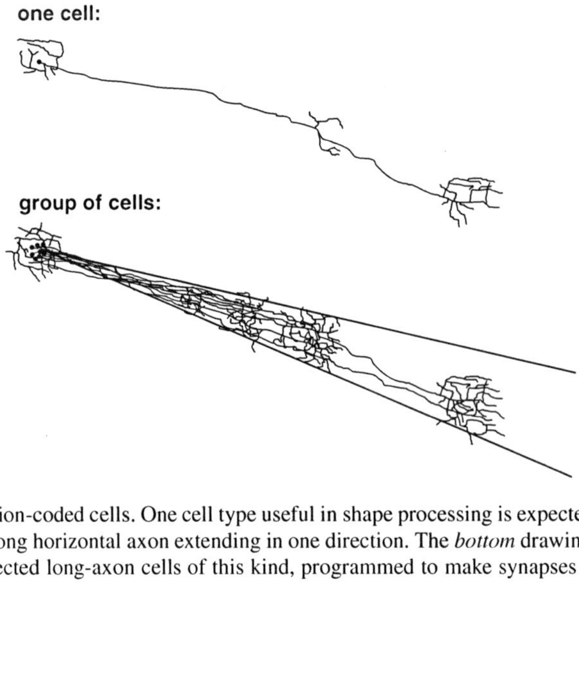

Fig. 9.1 Direction-coded cells. One cell type useful in shape processing is expected to be the layer 2/3 cell with a long horizontal axon extending in one direction. The *bottom* drawing shows a group of similarly directed long-axon cells of this kind, programmed to make synapses within a narrow angular range

9.2 Direction-Coded Cells

The next few drawings are meant to be in an idealized cortical top view which flattens out the cortex, superimposes the cortical map on the retinal map, and ignores changes in cortical magnification. Some of the long fibers shown go down into white matter and re-emerge to arborize, but this, too, is ignored in the drawings.

The neurons showing the selectivity of interest are probably layer 2/3 neurons of the visual cortex, known to have a primary and a secondary receptive field, where

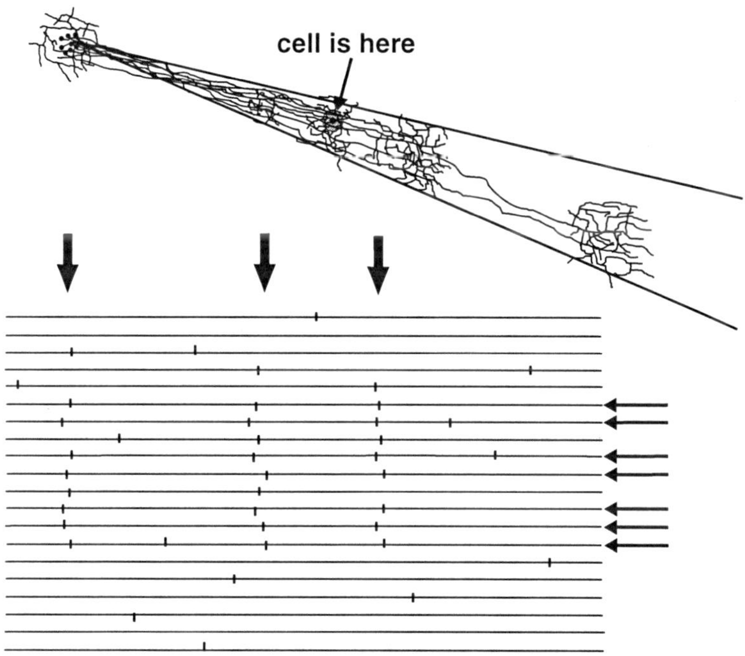

Fig. 9.2 A cell receiving volleys from direction-coded cells. If the cell indicated in the *top drawing* can recognize different direction-coded cells by their chemical signature then it can in effect recognize the *direction* from which the volleys shown in the *bottom drawing* arrive to it

the secondary (non-local) field tends to contribute a subliminal effect (Gilbert *et al*, 1996; see review by Seriès *et al*, 2003). The generally assumed arrangement is that some cells having cell bodies at the secondary receptive field have long horizontal axons, and make their contribution through synapses on cells which have their cell bodies at the primary receptive field.

A useful (and equivalent) description of the arrangement states that every such cell has, besides its receptive field, an *effective field*, to which its axon sends its outputs, and throughout which it contributes to the secondary receptive field of cells there. The *assumption of reciprocal connectivity* (where valid) states that the effective field of such cells is lined up with their secondary receptive field (or fields).

The direction-coded cells of Fig. 9.1, with axons confined to small sectors, may be the small pyramidal cells (sometimes described in older papers as *granule cells*) known to exist in layer 2/3 of the striate cortex.

Since the axons in Fig. 9.2 have their branching confined to a narrow angular sector as shown, the cell whose input is shown in the traces (bottom) has to be in that sector, in order to be reached by their volleys. (It is assumed that the local branches of the axons are distinguishable from the long branches by their signature.)

9.3 Relational Firing: Two Cell Groups Broadcasting a Relation

The concept of direction-coded cells can be taken one step further; one can envision two groups of direction-coded cells, so positioned that they reach one another, as linking up and performing simultaneous ignitions. By doing so, these cells can broadcast their relative direction as illustrated in Fig. 9.3.

The arrangement can be considered as an example of the more general idea of "relational firing" by which two groups of cells, both with labeled axons, coordinate their activity (mutually or in one direction) and thereby, through labeled lines, tell all cells they reach (including cells in other cortical regions) that they bear the relation written in their labels.

It is noted that synchrony can be envisioned without *direct* reaching; however the linkup protocols in Chapters 19 and 20 see to it that the reaching in the cases of repeated co-ignitions is direct.

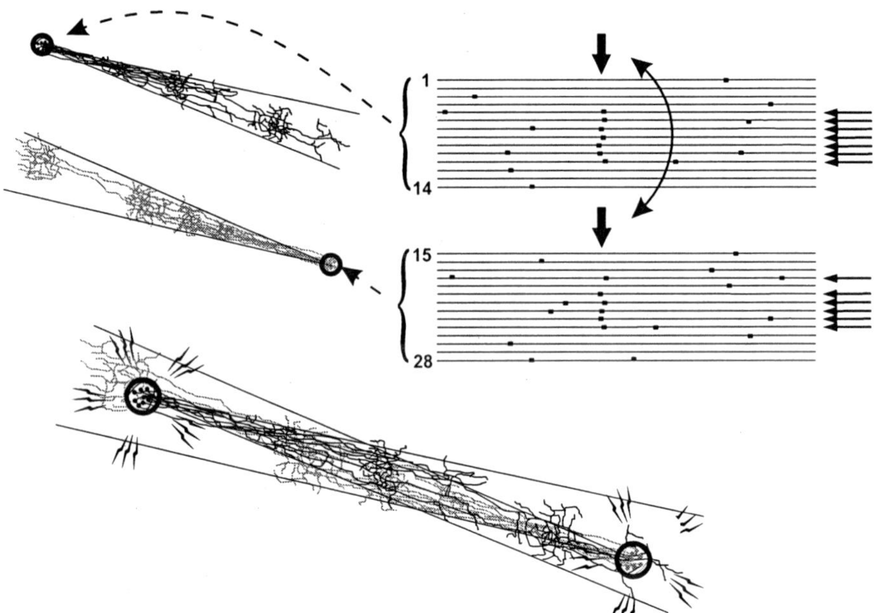

Fig. 9.3 Two cell groups broadcasting their relative direction. If the direction-coded cells can communicate relative direction through their chemical signature, it follows that by synchronizing their volleys, two opposite-directed cell groups can broadcast their relative direction. The fact that the first group is able to follow the volleys of the second group indicates that its cells are reached by the axons of the second group, and (by Fig. 9.2) that its cell bodies bear the indicated directional relation to those of the second group; the same conclusion, the other way round is drawn from the fact that the second group is able follow the first. Spike trains 1 through 14 belong to one of the cell groups; spike trains 15 through 28 to the other; the two cell groups, superimposed, are shown at the *bottom*

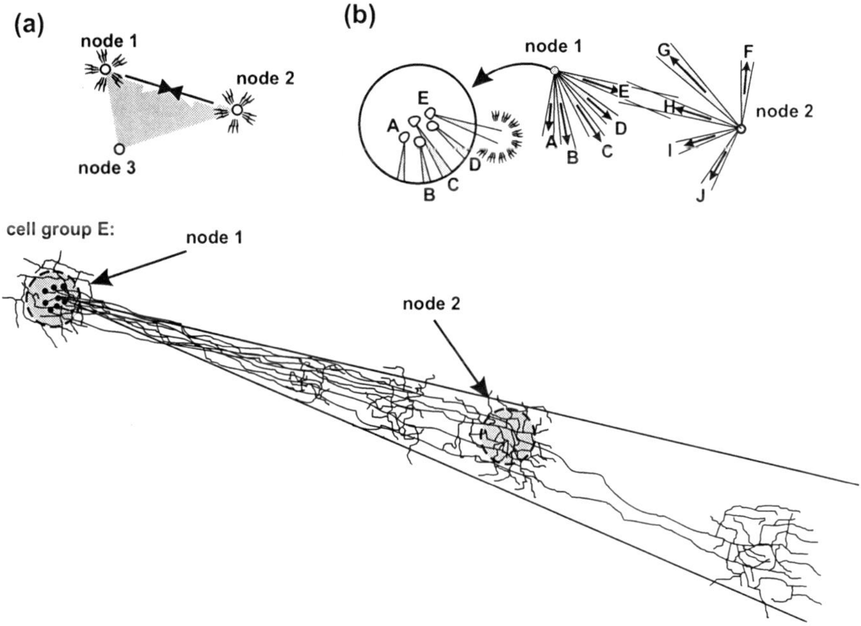

Fig. 9.4 Each cortical locality has many groups of direction-coded cells. Every cortical locality which may need to communicate its relative direction with respect to another cortical locality (represented by one or more cortical hypercolumns, Sects. 6.3 and 15.5) contains a supply of direction-coded cell groups, ideally enough to reach all possible other cortical localities. In the *triangular shape* at (**a**), the cell group E at one corner is properly positioned to link up with cell group H at the other (see (**b**))

As a comment on *labeled lines* signaling, it may be mentioned that in the case of Fig. 9.3 each cell group only needs to recognize that the other cell group has directional coding equal and opposite to its own; the cell groups in this situation do not need to be able to recognize all possible directional codes.

9.4 Visual Sentences Conveying that Two Sides Meet in One Point

As may be anticipated, what I am driving at is that all shapes can be approximated by polygons, and polygonal shapes can be described by means of directional co-ignitions; for instance a triangular shape can be described by three-directional co-ignitions, each involving a conjugate pair of direction-coded cell groups, as shown in Fig. 9.5. However, a more careful look at Fig. 9.5 shows that one thing is missing. Each corner hosts two direction-coded ignitions, and it is necessary to convey in some way that in each case the two are made up of cells with cell bodies in the

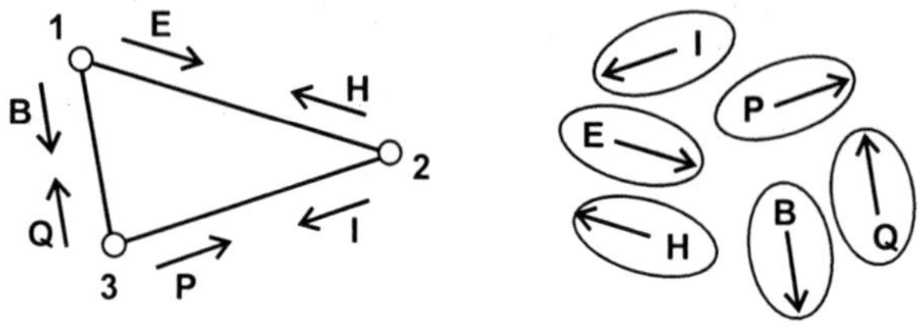

Fig. 9.5 Direction-coded cell groups used in describing a triangular shape

same cortical location. It amounts to the step of joining pairs of sentences on shared nouns (Fig. 8.1).

The challenge here is that the two ignitions involving the same corner of the triangle generally involve two different sets of cells. This is clear in Fig. 9.4(b), where for instance the cell groups E and B in location 1 (whose letter designations match this figure and the next) are clearly distinct. If ignitions are to be identified by the cells that participate in them, the ignitions of E and B will appear unrelated; the direction-coding-based labeling of the cells will supply no clue that the ignitions involve the same cortical locality.

It could be suggested that, because of retinotopic mapping, it is "obvious" that two sets of cells with cell bodies intermingled in the cortex are "known" to the system to be in the same place, and therefore there is no need for any special message to convey the fact that the cells B and the cells E are close together. However, in the later stages of processing, cell input locations and cell body locations become less strictly linked; cells will often receive significant input from two or more cell groups located far apart. The retinotopy-based relations between cell groups must be cast in a form of mutual timing at an early stage of the processing, where they are not yet lost, namely at the level of the primary visual cortex.

9.5 "Kernel Cells," Used in Joining Co-ignitions on Shared Points

To identify location through synchronized ignitions, it is necessary to add another cell type, to be called *kernel cells*, with the property that they only co-ignite with cells whose cell bodies are in the immediate vicinity of their own. This will then ensure that when two cell groups both co-ignite with the same kernel cells, they broadcast, by doing so, that the two groups have cell bodies close together.

To have the required properties, a kernel cell is not allowed to have a long intracortical axon, and must in addition receive its input exclusively from the *local branches* of axons of the direction-coded cells, the branches making synapses near the cell body.

The ignitions shown in Fig. 9.6 illustrate the way co-ignitions of kernel cells can convey the fact that the groups E and B have cell bodies in the same place in the cortex.

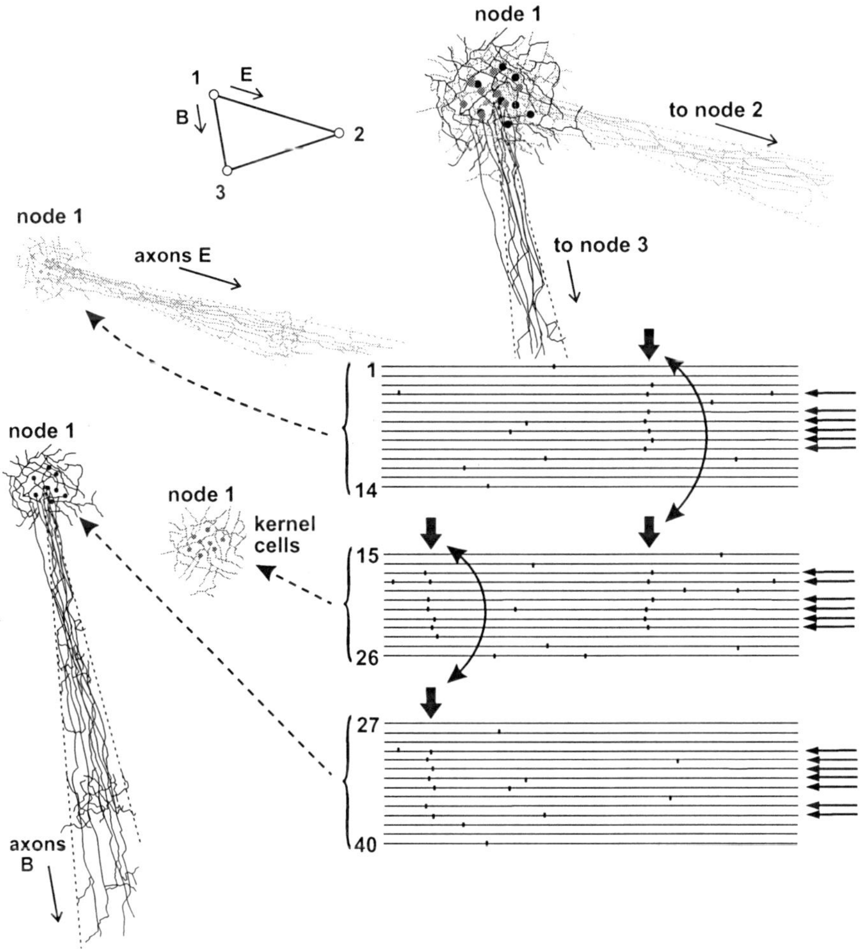

Fig. 9.6 Broadcasting that two direction-coded groups are at the same place. One co-ignition broadcasts that one direction-coded group has its cell bodies at the same location as a certain set of kernel cells, and another that a second direction-coded group has them at the same location as the same set of kernel cells. The two messages together say that the two direction-coded groups are in the same location. It will be noted that the two ignitions of kernel cells can be identified by being the same through the fact that they are made up of (largely) the same cells (*arrows at right*)

Like the direction-coded cells, the kernel cells communicate their message through *labeled lines*. Instead of being direction-coded, the kernel cells are "proximity-coded cells"; both kinds of cells respond exclusively to volleys from cells having cell bodies in a certain spatial relation to their own.

In one case the built-in growth instructions make the intracortical axons grow in a specified direction, in the other case they make them grow short. In both cases the molecules carrying the instructions are present in (or around) the same axons which must later use the information when telling the neurons receiving spikes from them of the nature of the cells originating the spikes. The chemicals which regulate the growth and connectivity of these cells are also in a position to serve as their signature, telling other cells what growth features they dictate to their own cells.

Toward the higher cortex, V2, the kernel cells at any given point convey the *identity* of the point (*node*). It must be emphasized that, despite their short intracortical axons, the kernel cells cannot be characterized as "short-axon cells," because their axons must go down into the white matter and then reach area V2, in order to broadcast the point identity there.

The co-ignitions of Fig. 9.6 are an example of relational firing between dissimilar cell groups. In this case the synaptic linkage between the groups is also not symmetrical; transmission is one-way from kernel cells to direction-coded cells. The relational firing principle is still satisfied; the direction-coded cells know that the kernel cells are near their cell body, and by co-igniting with them they broadcast the spatial proximity.

It is in principle enough for the kernel to co-ignite with the local directional cell groups one at a time, but that would mean that the kernel cell's ignition rate would be much higher than that of the directional groups, especially at localities where many directional groups have links to other localities (Fig. 9.11). It is more efficient to schedule a new kind of co-ignition, in which a kernel ignites together with all directional groups on its node, all at once; at least all the ones free to ignite at the moment.

I will call new kernel co-ignitions "*vertical* ignitions" (even though the directional groups are probably in the corresponding *orientation columns*, and *horizontally* displaced from each other by slight amounts). The latter arrangement requires that the kernel have ignition-transmitting contacts to all the local direction-coded groups, therefore the direction-coded groups must passively follow the kernel groups, but not vice versa. To distinguish, I will use the term "*horizontal* ignitions" for the co-ignitions, like directional co-ignitions, which use long horizontal fibers.

9.6 Visual Sentences Communicating a Triangular Shape

Using the co-ignitions just described, a whole triangle can be conveyed by a series of co-ignitions, as shown in Fig. 9.7.

Each group of traces in Fig. 9.7 is made up of members of a different "neuron pool," some being kernel cells, others directional cells divided into pairs of conjugates. The set of traces 1 through 14 is repeated at the end of the sequence for illustration of its alignment with the traces of the right-hand column, and the ignitions A and F appear in both places. As in Fig. 9.3, the axonal arborages of the neuron sets are shown at the left of the traces (this time drawn smaller, to save space).

The co-ignitions involving kernel cells, B for corner 1, F for corner 2, and D for corner 3 (arrows on top), are shown here as triple ignitions; in practice as many of the co-located directional groups are expected to ignite together with the kernel cells as are recovered from their previous ignition when the kernel ignites (Sect. 16.2).

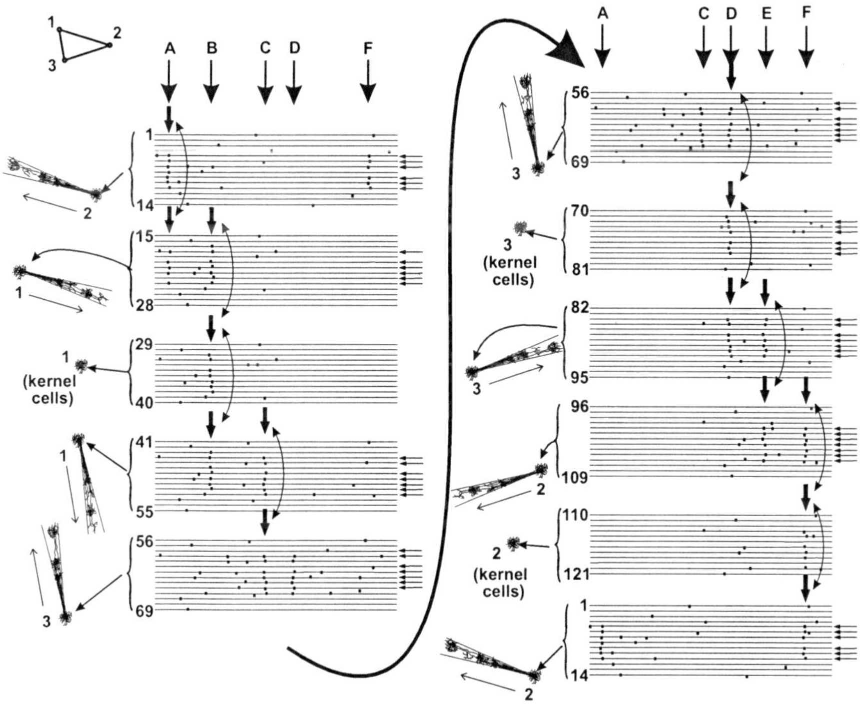

Fig. 9.7 Co-ignitions describing a *triangular shape*. Three pairs of directional ignitions joined on three shared corners can broadcast a triangular shape. To show this, it is only necessary to repeat the idea sketched in Figs. 9.3 and 9.6 for all three sides and all three corners of the triangle

As indicated by the numbering of the schematic spike trains, every trace in the drawing involves a separate neuron, altogether 121 neurons in this sketch. It is not necessary for any one neuron in V2 to be reached by all of them; the purpose of the illustration is to demonstrate that the set of spike trains, together, contains all the information needed to reproduce the triangular shape in question.

The neurons together are supposed to send the triangular shape "upstairs" to V2. To do this they need to have axons going to V2. As it happens, the cells sending their axons to higher cortex tend to be in layer 2/3. The cells having long horizontal axons tend to be in the same layers, which neatly fits with the role ascribed to them in this model.

9.7 "Contour Cells" and "Direction-Coded Cells"

The two kinds of co-ignitions appearing in the last figures, one kind linking opposing direction-coded cell groups, the other linking direction-coded cell groups with their local kernel groups, are sufficient, as was seen, for transmitting triangular shapes. However, it will be noted that in order to transmit a polygonal shape with more than three sides it is necessary to transmit the distinction between the sides of the polygon (visible lines) and the diagonals. In the case of a quadrangle (Fig. 9.9), there is need for at least one diagonal, because the directions of the four sides alone will not, for instance, show a difference between a square and a rectangle. Polygonal shapes with more than three corners must be divided up into triangles, as is done in surveying.

Further, the "point skeleton" of an object (bottom left drawing in Fig. 9.9) needs to be transmitted in its entirety by means of direction-coded cells, to specify the relative directions of all point pairs, while the contour lines need to be transmitted by means of cells able to convey things like color opponency, and co-ignitions able to transmit which pairs of points are linked by segments of visible contour.

In keeping with the labeled lines concept, it is reasonable that the difference between sides and diagonals should be transmitted through the use of two different cell types, one used for describing the visible lines (contours) and the other for describing the relative directions between significant points.

I will use the term "contour cells" for the cells dealing with the visible lines and continue to use the term "direction-coded cells" for the ones dealing with rela-

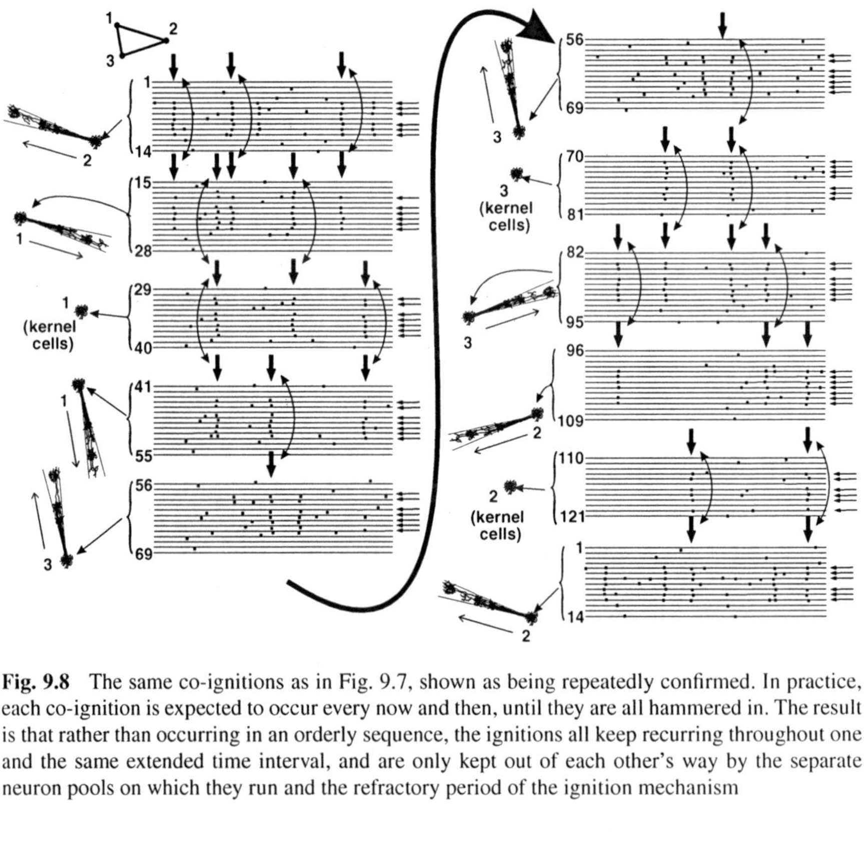

Fig. 9.8 The same co-ignitions as in Fig. 9.7, shown as being repeatedly confirmed. In practice, each co-ignition is expected to occur every now and then, until they are all hammered in. The result is that rather than occurring in an orderly sequence, the ignitions all keep recurring throughout one and the same extended time interval, and are only kept out of each other's way by the separate neuron pools on which they run and the refractory period of the ignition mechanism

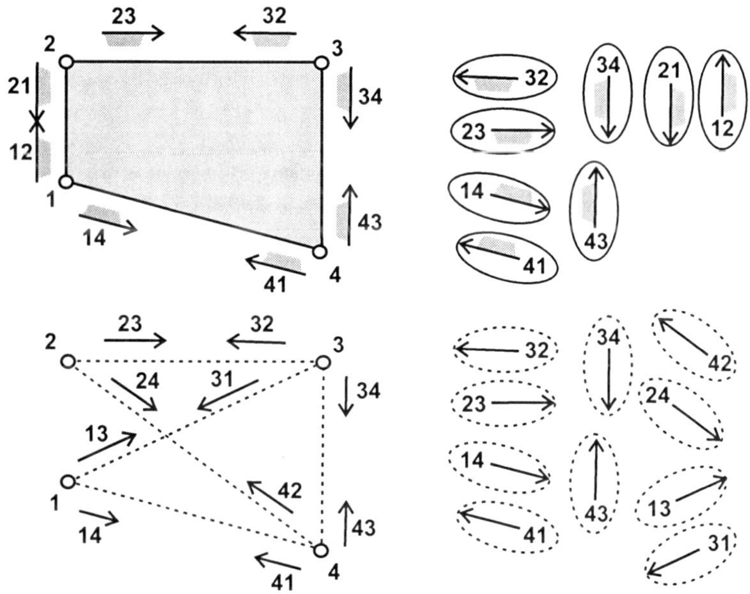

Fig. 9.9 Cell groups for the contour and point skeleton of a quadrangle. To transmit a four-cornered shape, four directional co-ignitions are not enough; it is necessary to add at least one diagonal. And since contour cells and direction-coded cells transmit different aspects of the shape, for instance directional co-ignitions do not reveal whether the line segment they describe is visible or is a diagonal, a separate set of co-ignitions is necessary for the contour (*top*) and the point skeleton (*bottom*), linked in each case on the corner points by means of kernel cells. The drawing does not include a directional co-ignition between points 1 and 2 to emphasize that unambiguous representation does not require every contour segment to be duplicated by a directional co-ignition

tive directions. *Contour cells* are expected to be specific to local orientation of the contour, as well as such properties as contrast of color and luminance, whereas the direction-coded cells, which often describe the direction of invisible lines such as diagonals, only need to be specific to the relative directions.

In direction-coded cells the long axons must grow in one direction (to make their synapses in the retinotopic region of one angular sector of real-world relative directions), otherwise their directional message is ambiguous. In contour cells it is advantageous for the axons to grow in a number of directions (which enables them to follow curved contours, see Chapter 19, but tends to make them unsuitable for conveying relative direction between points), and it is necessary to retain a preference between either one or the other of the two opposing directions along the preferred orientation (Sect. 19.5).

It can be assumed that the cells described in the paper of Gilbert and Wiesel (1983) as having long horizontal axons growing in a number of directions are contour cells, except when they have a completely symmetrical system of long axonal branches, in which case they are outside the cell types included in this book.

To my knowledge, a class of cells matching the description of direction-coded cells (Fig. 9.1) has not been studied or specifically sought out up until now; however, I predict that once they are, these cells will show themselves to be a very populous class.

Because contour cells are generally unsuitable for conveying relative direction, transmission of a polygonal shape must contain enough directional co-ignitions to describe the full point skeleton of the shape. This means that the visible lines are often transmitted by direction-coded cells as well as contour cells.

It can be assumed that the transmission of shapes is in practice usually overdefined, meaning that more than the minimum required number of point pairs is usually transmitted; at the same time, since for many-cornered shapes the number of point pairs increases approximately quadratically with the number of corners, and reaches fairly large numbers, it can also be assumed that for them not all point pairs are transmitted. The transmission is unambiguous as long as *enough* point pairs are transmitted by directional co-ignitions that every point can be tied to the others via triangulation.

As will be seen in Chapters 19 and 20, neither contour cells nor direction-coded cells are expected to fall into any of the classical receptive field categories, as they (as well as the kernel cells) play their roles in the next functional layer of processing. They are expected, for instance, to perform repeated self-ignitions in certain phases of their operation.

All the same, Ts'o *et al* (1986) clearly saw layer 2/3 cells with long horizontal axons, and recorded their co-ignitions in cross-correlograms. They described their cells as having classical-appearing receptive fields, and reported that the preferred orientations of correlated cells tended to agree with one another. It is possible that the recorded cells included both contour cells and direction-coded cells, and that under anesthesia the direction-coded cells show their encoded relative directions as *preferred orientations* and show the horizontal extent of their axonal arborages as *receptive field sizes*.

9.8 Broadcasting More Complex Shapes

We know from geometry that the slopes of three sides of a triangle determine the triangle, except for its size and location. In the same way, three pairs of direction-coded cell groups determine it, except for size and location, as was seen in Figs. 9.7 and 9.8.

The power of shape transmission through points and directions comes from the fact that all plane figures can be approximated by grids of points, and then turned into grids of triangles (polygon mashes) by means of straight lines drawn between the points; this means that enough triangles, joined on their shared corners, can describe any plane figure, except for size and location.

In the context of a model aiming to describe shape processing in the visual system, the inability to capture size and location is not a drawback, and in fact adds to the realism of the model, since vision is also insensitive to size and location.

It must be added, though, that the present model is highly sensitive to rotation of the whole shape, and the objection may be raised that it should be insensitive to it. However, the rotation-insensitivity of actual vision is in fact quite limited. (If vision were rotation-insensitive, reading the label on a rotating phonograph record would be easy, but in fact it is not.) In the present model we assume that rotation is processed along similar lines to three-dimensional rotation (which results in perspective distortion), and that they can both be captured by tracking (Fig. 16.2 and Sect. 16.4); a more detailed treatment is beyond the scope of the model.

If one triangle can be broadcast by means of points and directions, and any number of triangles can be joined together on their shared points, the conclusion is that all grids can be broadcast by using the same simple visual sentences. For instance, Fig. 9.10 shows the schematic resolution of a slightly more complex plane figure to the two kinds of co-ignitions.

It can be suggested that the sets of locations (the sets of *nodes*, Chapter 15) supporting the two kinds of cells do not need to be identical; however, as will be

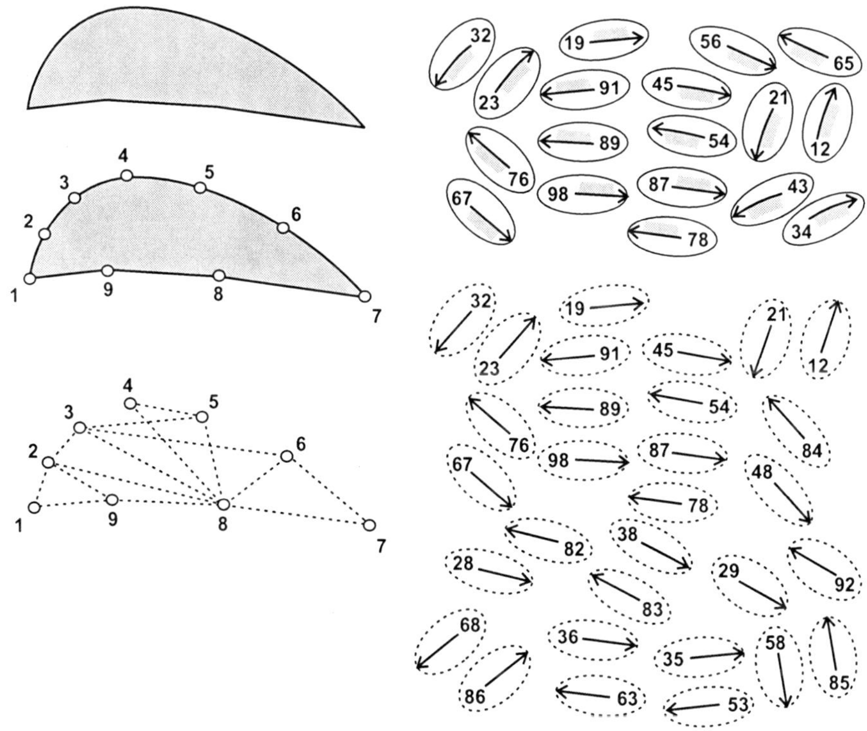

Fig. 9.10 More complex shapes may require very many co-ignitions. The idea illustrated in Fig. 9.8 can be extended to shapes with many corners; the difference is that when there are many corners the point skeleton requires many more co-ignitions than the contour. Since the shape has some *curved contours*, the *arrows* describing some of the contour cell groups (*top right*) are drawn curved

seen below (Part IV), the first linkup between nodes, which involves "crawling" along the (often curved) contour sections between nodes, is a creation of contour links. The contour links are the first hardware records of the discovery that two nodes are on the same contour. The directional links, some of which are between non-adjacent nodes, build on the fact that the nodes linked by contour links are on the same contour. Therefore, at least initially, the two sets of nodes must be identical.

The number of links meeting in any point in the grid equals the number of sentences to be joined on the noun corresponding to the point, or equivalently the number of directional ignitions, plus the number of contour ignitions, to be joined by use of the kernel cells at the point. Figure 9.11 shows the way some the corresponding co-ignitions may look.

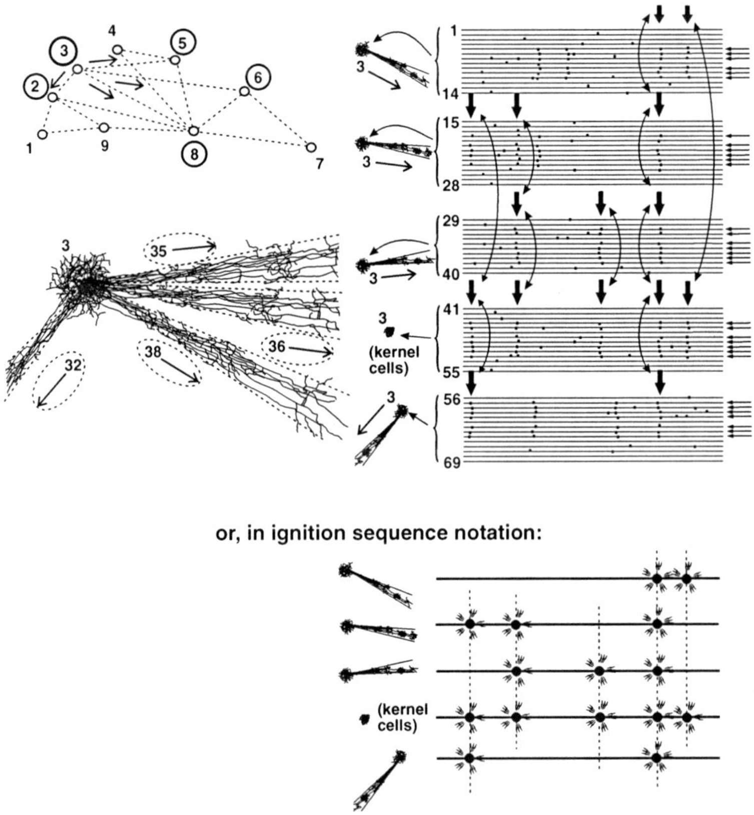

Fig. 9.11 Joining several links on one corner point. The figure singles out one corner, corner 3, in the point skeleton of the shape shown in Figs. 9.10, and shows how the four directional links converging there can be joined. Only the co-ignitions of directional cells with the kernel cells are shown here; when describing the whole shape, the two contour ignitions involving corner 3 (ignitions 3 with 4 and 3 with 2) should also be joined to the rest. At the *bottom*, the ignitions of the multiple join are also shown in the more condensed "ignition sequence notation"

It will be noted that some of the co-ignitions shown in the traces at the right involve more than two of the links; one of them involves all of them. All co-ignitions involve the kernel cells.

As seen in Fig. 9.1, the effective fields of individual direction-coded cells do not fill the entire "wedge" of visual field filled by the whole cell group, but only a few-hundred-micron neighborhood, or a few of them. This, as will be seen later, is useful in avoiding ambiguity during linkup in the cases where several of the points of interest are lined up on a straight line.

For the purposes of the messages received "upstairs" (in V2), the distance between the cell body and the synapses making the contacts is of no interest; the upstairs is (in this model) insensitive to distance, and is sensitive only to relative direction. However, in the processing "downstairs" (in V1), where it is decided which points are to be linked to co-ignitions, a distinction must be made between pairs of locations lined up on the same straight line. All co-ignitions must involve two points at a time; otherwise the subsequent joining operations will yield indeterminate results.

9.9 The Role of Retinotopy and Connectivity

The ability of the cortex to translate distinguishable relative directions into co-ignitions results from a favorable confluence of two properties of the cortical network.

The first is that there is, by and large, a one-to-one mapping of retinal points to cortical columns, and the grain size of the mapping is roughly the distance between columns (or the distance between CO blobs), in the sense that two *phosphenes* created by cortical microstimulation only appear distinct (by and large) when the stimulation sites are farther apart than the columns (Bak *et al*, 1990; Schmidt *et al*, 1996).

The second is that the cortical columns are both small enough that the neuron groups in them are within the reach of each other's local axonal branches, and large enough (or almost large enough, Chapter 17) that the same neuron groups reach the corresponding neuron groups in other columns. If there were no reaching *within* columns, the linkup between direction-coded groups and kernel groups, needed to support the joining of co-ignitions, would not be possible, and if there were no reaching *between* columns (or small groups of columns; Sect. 17.1), the connections between pairs of facing direction-coded groups, also needed to support the co-ignitions, could not be created (Chapter 19).

In this connection it will be noted that the distortion of the retinotopic map does not matter, as long as neighborhood relations are preserved. In reality the retina is spherical, the cortex is folded, the "horizontal" axons often go through white matter and make a shortcut through a gyrus, and the cortical magnification is different at different distances from the center; yet all this does not interfere with the ability of the cells to be coded for relative direction, with a coding rooted in objective reality, in this case the outside-world spatial relations.

This is why it does not matter that in the foregoing sequence of drawings (as well as the drawings below) the nonlinearity is ignored and the cortical sheet is represented as being flattened and uniform.

9.10 Confirmation Loops and the Epochs on High Poisson Surprise

The spike trains shown in Fig. 9.8, which shows the way ignitions are hammered in through repetition, can be placed side-by-side with the observation that during vision (many) neurons of the visual cortex show occasional epochs of elevated spike rate. As was noted, the spike rates during these epochs are not necessarily very high, but high enough to make some of the epochs highly surprising (Fig. 9.12).

Two mechanisms are expected to cooperate in raising the spike rate: one is the confirmation loop, and the other is the need to join sentences on their shared nouns. For instance in Fig. 9.8, the neuron numbered 85, which is part of location 3, participates in four ignitions during the time-interval shown. Two of them are co-ignitions with the local kernel cells and two are co-ignitions with conjugate cells of location 2. The two kinds of co-ignition sentences are joined through the fact that the same neurons participate in them (in the illustration these are shown as traces 84, 85, 87, 88, 89, 90, 91, singled out by the arrows at the right).

The repeated co-ignitions, in each context separately, are caused by mechanisms distinct from the ones causing the random background firing of the same neurons; therefore they disobey the probability distribution governing the random firing. It is noted that the firing during the bursts is irregular. This is consistent with the fact that single-unit records only show the behavior of one member of each synchronized cell group; and the timing of the volleys often depends on events involving other members.

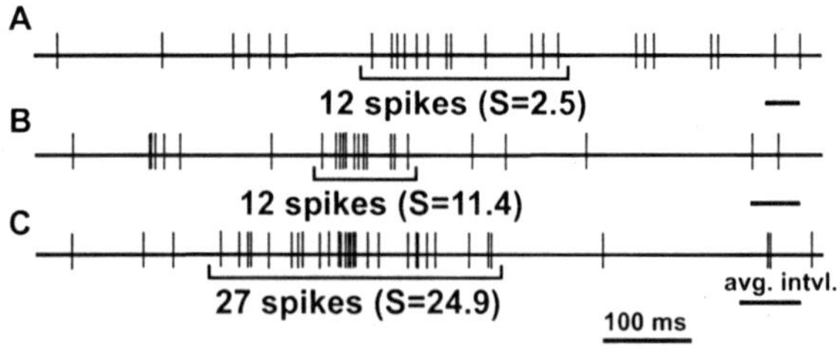

Fig. 9.12 Joined repeated ignitions show up as bursts in the visual cortex. The figure shows three records taken from a neuron in the visual cortex of a cat, while the animal was freely looking around in the lab. The records were subsequently scanned off-line for epochs of high Poisson surprise (Sect. 3.3). The surprise (S), defined as the negative logarithm, base 10, of the Poisson probability of accidental occurrence, is shown with each of the bracketed time intervals (redrawn from Legéndy and Salcman, 1985)

Even though records like Fig. 9.12 are taken from single neurons, and therefore miss the surprise of simultaneous spiking by neuron groups, they are (often) able to detect the biological effort to convey surprise thanks to the extra spikes forced into the spike train by confirmation loops; hence the usefulness of Poisson surprise runs in diagnostics.

9.11 3D Extension of Polygon Graphics

The polygon graphics concept, as described in terms of directional relations, extends in a natural way to three dimensions.

The directional network builds up its grid on the directional vectors along a set of straight lines in real-world space, and only needs to keep track of the identities of points in which the lines meet. The stable building blocks of shapes, which are *triangles* in two dimensions, become *tetrahedrons* in three dimensions.

It may be noted in this connection that in *projective geometry*, the discipline which formalizes the art of drawing objects in perspective, straight lines always remain straight and the intersection points between lines are always preserved.

The binocular inputs to simple and complex cells, serving as the basis for stereoscopy, have been discovered very early by Hubel and Wiesel (1962), and an extension to three-dimensional contour cells and direction-coded cells appears natural.

However, for the sake of simplicity, the descriptions below will remain two-dimensional.

9.12 Distributed Knowledge

One twist to the "knowledge" concept, as applied here, is that the knowledge transmitted to the upstairs by a neuronal network is made up of several diverse pieces, and the pieces are carried by distinct sets of neurons, none of which knows all the facts contained in the whole cooperative action.

The counterintuitive part of all this is that the next layer of processing also does not need to contain any single neuron that knows all details of the picture. For instance no neuron needs to be reached by all 121 spike trains shown in Figs. 9.7 and 9.8. The whole picture still remains a set of details linked together, although the distribution of component items among the neurons changes.

This is why there is no "grandmother neuron," or any need for one. How this is possible is not clear from the foregoing abbreviated remarks, but it is hoped that the examples described in the rest of this monograph contain enough detail to demonstrate that processing based on distributed knowledge is possible.

Part II
Contour Strings and the Contour Wave

Chapter 10
Enter the Contour String

The retinal image of a typical natural scene is a tangle of shapes, and it is up to the visual system to divide it up into entities which correspond to objects worth dealing with.

It is likely that the visual system employs a number of strategies to deal with the challenge. This writing explores only one such strategy, which is, in effect, to look for extended contours in the visual field, long enough to contain many elementary receptive fields lined up end-to-end, and utilize the surprise inherent in simultaneous stimulation of such aligned receptive fields to generate surprising events.

The rationale is that a contour is likely to be the boundary of some object. Or, said differently, it is a safe guess that the cells responding to the contour all respond to the same real-world entity.

10.1 The Issue of Enabling Communication Between Parts of an Image

The idea that time-linked ignitions can broadcast shapes, as described in Chapter 9, raises the issue that before the neuron groups can ignite together, they must first link up, and before they can do that, they must find each other. The question is how do they manage to do that? After all, the neuron groups in question may be quite far apart in the cortex, and no functional linkage exists between them to start with. The contour does not have any tangible physical reality; its only manifestation is that certain cortical cells, independently, happen to be in the cortical image of the same contour!

In order to give them a basis for recognizing that they are on the same contour, every contour must first give rise to *an avenue of activity* inside the cortex, temporarily able to channel information back and forth between neuron groups along its path. Creating an avenue of activity is the subject of the next series of sections; the task of turning it into an *avenue of communication* is taken up in Chapter 19.

C.R. Legéndy, *Circuits in the Brain*, DOI 10.1007/978-0-387-88849-1_10,

10.2 Cells Which Link Up to Pass Waves When Co-stimulated

To detect a contour, it is necessary to find a protocol that interlinks simultaneously stimulated simple cells having similar receptive field properties, including preferred orientation and color opponency, and having, in addition, receptive fields lined up approximately collinearly.

If the brain can manage to identify such a set of cells, and make its members perform in a cooperative manner, it can be said with great confidence that the resulting cooperation is not caused by noise but signifies the presence of a contour in the world outside.

10.3 Ignition on the Contour Moves Like a Wave (Contour Wave)

If the interconnected cells have linkages strong enough to pass firing between the cells, the group will become capable of cooperative firing, very much like ignition.

More precisely, since the firing spreads in a chain-like fashion, the "chain ignition" of a contour string is not really like an "ignition," which would be characterized by nearly simultaneous firing of the participating neurons, but is more along the lines of a disturbance propagating along a line, like the serial tipping of dominoes. I will call it a "contour wave."

It is worth emphasizing that in the present usage the term *wave* is not meant to describe a sinusoidal disturbance like a sound wave or an electromagnetic wave, but a nonlinear pulse-like moving configuration like the nerve impulse or the "domino wave." Accordingly it is permissible to speak of "one contour wave" or several of them, depending on the number of pulses moving after one another.

Unlike the dominoes, which only interact with their nearest neighbors (as the cells in the simplified Fig. 10.1), the neurons in the chain ignition receive inputs

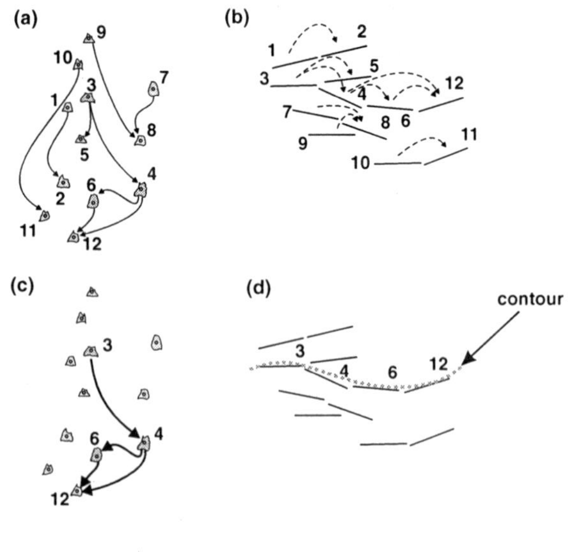

Fig. 10.1 The concept of chain excitation enabled by co-illumination. The figure illustrates the idea which gives rise to contour waves (leaving out many details). (**a**), (**b**) Simple cells with similar and aligned receptive fields are connected (from left to right in the drawing) by links which lie dormant most of the time but become active when the cells they connect are both stimulated from their receptive fields (**c**), (**d**) Appearance of a contour activates a subset of the dormant links

from a number of their neighbors upstream of them (Fig. 10.3), as is made necessary by the surprise requirement of firing.

In the case of the contour string, this same multi-input requirement serves as a part of the built-in test which makes it unlikely that the chain ignition can propagate as a result of a series of chance coincidences of firing, without any contour being present.

In order to avoid reverberation, it is also necessary that the participating cells only contribute a single spike to each cooperative event. The data processing advantage of avoiding reverberation is not confined to near-simultaneous ignitions; it extends to the contour waves.

10.4 Seeing Viewed as Short-Term Learning

The concept of the stimulus-induced linkup between cells is an example of what is a recurring theme throughout this monograph: that the processing of visual images involves *synapse modification* of the sort usually described in connection with *learning*.

The effects of the synapse modification generally disappear fairly fast – in minutes, seconds, or milliseconds – not necessarily because of decay, but because the information is used fast, and is soon overwritten with other information.

A corollary is that, if vision indeed works as described in this model, the visual cortex can become a valid experimental proving ground for the study of *learning* at the neuronal level.

10.5 Only the Simple Cell Is Suitable for Conducting the Contour Waves

It is important that only collinear cells combine into a contour string, so their cooperative activity carries some assurance that the underlying contour is a single contour, rather than for instance two contours close together.

From what we know of simple cells, one can state with confidence that the simple cell is the proper cell type for the purpose, because the complex cells may make the same response to a slightly shifted version of the contour, and in that way the interconnected cells are no longer guaranteed to belong to a single visible object, as illustrated in Fig 10.2.

10.6 The Need for Drome-Selectivity in Simple Cells

The contour waves, in order to function properly, require in addition that all their participating simple cells have a built-in propagation drome, as in Fig. 10.1. In other words, any given simple cell must either receive inputs only from the left and send outputs only to the right, or vice versa; and the cells must only link up with cells having the same drome.

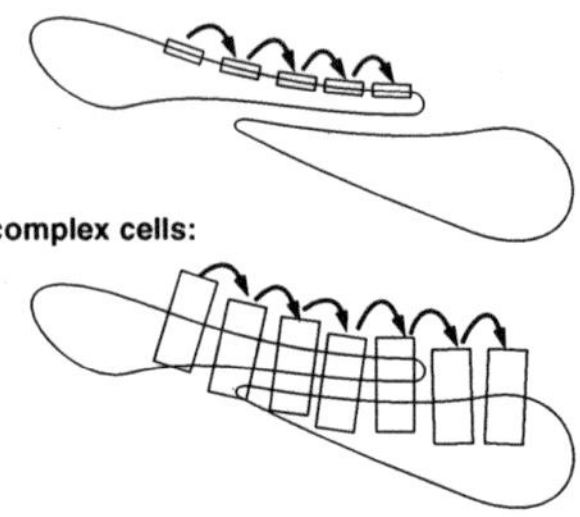

Fig. 10.2 A wave of complex cells failing to adhere to a contour. Two shapes (*shaded*) are so drawn that some of their contours run parallel over some part of their length, and present nearly the same orientation and color contrast while being parallel-shifted relative to one another. If the cells of the contour string were complex cells (*bottom*), insensitive to such parallel shift, the string would sometimes hop from one contour to the other, and defeat the purpose of the "same-contour detector." If they are simple cells (*top*), the contour wave (ideally) adheres to one contour

The reason is that otherwise two waves could move in opposite directions on the same set of neurons, and upon colliding would both die. The refractoriness of the shared cells would prevent any further propagation in either direction.

To avoid collisions, and the resulting disappearance of contour waves, the waves in opposite directions must use separate tracks, as it were, meaning that, in their ability to link up and conduct waves, the simple cells must be drome selective.

10.7 The Problem of Converting "Facts" into "Events"

One functional aspect of the contour string concept is that contour strings can generate surprising events; a feature whose significance may be underlined by pointing out the error in the simpler concept, sometimes implicit in older discussions of vision, that the visual system combines parts of an image into a whole simply through the fact that individual neurons respond to various details of the image.

The suggestion is that these neurons will respond together whenever the object appears, and therefore a scheme along the lines of Hebbian synapse modification can simply tie the neurons together to make them, for instance, into an ignitable group.

Such a notion assumes that, just by passively following their input from the lateral geniculate nucleus (LGN), the neurons responsive to various details of a picture (in our case a contour) will respond in a manner well enough coordinated that their correlation is detectable to other neurons.

The reason this will not work is that the presence of a feature is what one may describe as being a "fact," or a "piece of truth", and as such it is steadily present. *It does not single out a moment in time at which a surprising event of firing is supposed to occur.*

In order to translate the simultaneous presence of details into a surprising multineuronal event, the brain must contain machinery for selecting a time when the event is to occur.

As will be seen in a moment, the contour string is an example of machinery that can "spontaneously" initiate coordinated events of firing at times chosen through its own internal workings.

10.8 The Contour String as a "Prime Mover"

It so happens that the simple expedient sketched in Fig. 10.1, chain linkage activated by co-illumination, and a few details added, can turn a string of simple cells into an "event generator" which can then act as "prime mover" for a network, and initiate coordinated acts of firing during a certain kind of stimulation.

The "events" in this case are spread out in time as the wave runs down along the contour, which will be utilized in continuity detection (Chapter 19), but are time-concentrated when viewed from narrow cortical localities, and can be utilized in telling the local neurons, by their repeated occurrence, when a contour is in their neighborhood.

As is illustrated in Fig. 10.3a, b, when the image of a contour falls on the retina, a chain of simple cells, with receptive fields matching the parameters of the contour, will become unstable and will repeatedly initiate propagating waves of firing for the duration of the stimulus.

The principle illustrated in this sequence is that, if simple cells are linked up in a chain, it will occasionally happen that the random firing of some cells close together in the chain sends enough spikes to a cell downstream of them to fire it; the newly firing cell in turn makes it even more likely that the cells downstream of it will also fire, and so on for every successive rank of cells. As a result, the firing will tend to run down from cell to cell along the line. The idea is similar to the one utilized in self-ignition (Fig. 1.10), except that here the connectivity is drome-specific.

All neurons in Fig. 10.3 show a baseline firing rate amounting to about 1–3 spikes over the time period of the traces. Whenever, by chance coincidence, three of five consecutive cells fire in close enough succession, the next cell in the chain receives most of the same spikes at the cell itself, plus the cell's output. The result is that the next neuron is almost certain to fire, and the one after that even more certain (at least in the idealized situation of the figure). As a result, neurons 7, 8, 9, and all the rest, will emit impulses in quick succession, and a contour wave will run down the line.

If one uses the parameters appearing in the figure, the maximum number of contour waves that can be initiated per second will be determined by the frequency with which the uncoordinated LGN bombardment can cause two successive neurons to fire via their chain links. Besides the probabilistic limitation, the rate is limited by the refractory period (not shown).

In the actual neuronal network, it is conceivable that the thresholds are slowly drifting along the same lines as shown in Fig. 7.1. When the spontaneous spike rate is too low to generate contour waves often enough, the internal parameters of cells are changed to raise the probability of spontaneous spike generation and contour wave initiation.

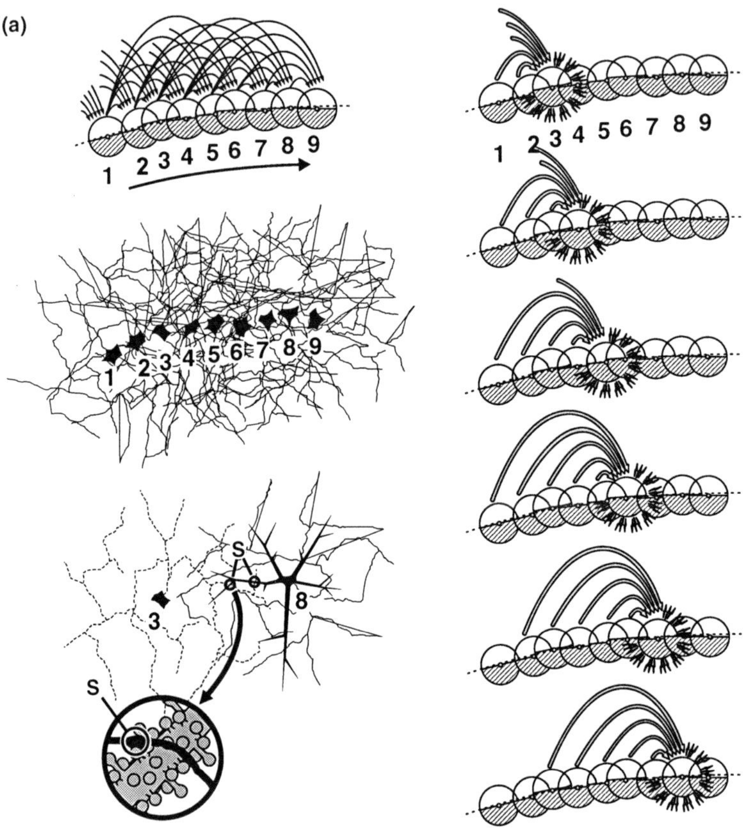

Fig. 10.3a Contour waves. Propagation of a contour wave. The receptive fields of nine simple cells are shown schematically, together with a segment of a contour (*broken line*) whose image falls upon the receptive fields and activates the dormant linkages between the neurons. The numbering of cells corresponds with the numbers on the traces shown in (b) below. *Arrows* between the receptive fields indicate the illumination-activated linkages. Each cell is shown as being contacted, in this case, by five other cells upstream of it, to be referred to as the "trigger set" of the cell (Figs. 12.1 and 12.2). The *bottom left* drawing, with magnified inset, shows roughly how the synaptic contact (S) between two of the cells ("3" and "8") may look. The linkages impose an ordering on the cells, to be followed (approximately) by the firing of the cells during contour wave action (sequence of drawings at *right*)

10.9 The Contour String as Representation of a Gestalt

The contour string is an example of a network which, as mentioned earlier, translates "connection" in the metaphorical sense into *connection* in the physical sense. I gave "contour strings" a separate name, other than for instance simply "contours," to emphasize that a *contour* is something in the outside world, whereas a *contour string* is a hardware device, custom-made to deal with a particular contour, and residing inside the neuronal network. It is the hardware representation of a very

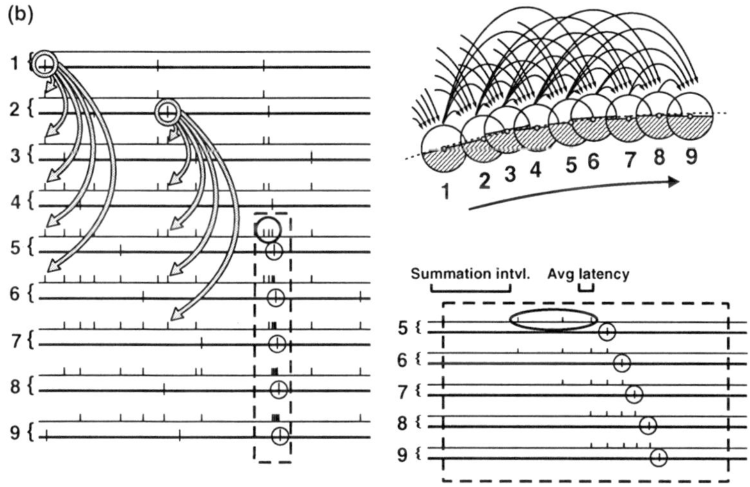

Fig. 10.3b Contour waves. Initiation of a contour wave by a random coincidence. Traces at *left and bottom*: input and output of cells 1–9. Two traces are shown for each cell. The *upper* trace represents the sum-total of input spikes received from the neurons of the trigger set (drawn to ignore conduction delay and the width of the unitary EPSPs); the lower trace for each cell shows the output. The trigger set of each neuron is made up of the five neurons preceding it in the chain; accordingly, all spikes emitted by a neuron show up in the input traces of the next five neurons of the chain (*curved arrows*). Threshold for firing is assumed throughout the drawing to be 3, meaning that when three or more members of the trigger set fire within one summation interval (*Summation intvl*), the neuron fires (after a brief delay, *avg latency*, which combines all conduction delays, external and internal). The portion framed with broken lines is time-expanded in the inset at the *bottom right*

simple *Gestalt* (von Ehrenfels, 1890; Wertheimer, 1923). In connection with the Gestalt concept, it may be added that the "Prägnanz" principle of Gestalt psychology is closely related to my *surprise* principle; they are two sides of the same coin. Both rely on the assertion that the world is not random, that things of interest to the brain have inherent structure, and that this structure in every case points the way to some (often simple) test able to lift the object of interest over the background.

In this way the contour string, with its contour waves running on it, can be viewed as a solution to the dual problem of selecting a moment in time to start cooperative events and coordinating the spikes emitted by cells far apart but subject to the same stimulation.

Chapter 11
Drift of the Retinal Image

A new layer of complexity is added, however, when one realizes that in actual living creatures the retinal image does not hold still.

In the case of the contour string and the simple cell, it becomes necessary to deal with the fact that, as a contour passes through the receptive field, the cell, initially dormant, must from one instant to the next become part of a contour string, perform its part in it, then after the contour leaves become dormant again.

11.1 *Tracking* the Nouns Used in Joining Sentences

We recall the central trick of the *polygon graphics* discussion in Chapter 9, which was the joining of visual sentences on their shared nouns (Sect. 8.2). One of the major challenges of *joining*, the drifting of the retinal image, was neglected in the earlier discussion.

Drift of the retinal image has the potential to ruin the whole idea of joining. When a co-ignition transmits the directional relation between point A and point B, and a little later another co-ignition transmits the directional relation between point B and point C, the upstairs may not recognize that the two "points B" are the same, because in the intervening time the node B has moved.

The identification is only possible if the node representing B can be *followed* in some way, from its location at the time of the first ignition to its location at the time of the second. In other words the *joining* of sentences on shared nouns, in the visual context, relies on successful following (*tracking*) of the noun objects.

The subject of tracking will be taken up in Chapter 13, and further elaborated in Chapters 16, 17, and 18. It will be seen that objects can be tracked, but the catch is that the image movement is not allowed to be too fast. When the image movement is so fast that tracking fails, the identifications will fail and shape perception will break down.

C.R. Legéndy, *Circuits in the Brain*, DOI 10.1007/978-0-387-88849-1_11,

11.2 The Word "Fixation" Is a Misnomer

It may be mistakenly suggested that the drifting of the retinal image presents no problem, because our eyes go through a series of "fixations," and during these periods they are stable enough to keep the contours on the same set of receptive fields.

However, as has been known for many years, the term "fixation" is a misnomer. Even in monkeys and humans, where the center of gaze dwells for a little while at selected places in the visual field, more accurate measurement reveals that during fixations the eyes drift around without any obvious plan.

More than that – it has been found that artificial stabilization of the retinal image interferes with pattern perception (Riggs and Ratliff, 1952).

Still, it is not immediately obvious just why. It is difficult to tear oneself away from the intuitive notion that ideally the eyes would like to make the image hold still, and only fail to do so because of some sort of imperfection in their feedback control system.

But that idea falls apart when one studies the eye movements of a cat. As it happens, cats do not have saccades at all, nor do they ever really "fixate." Their eyes simply move around, sometimes more lazily, at other times more hurriedly!

I became aware of this in the 1980 s, in the course of my studies of the cortical cells of cats during periods when they looked around freely in the lab. It was necessary to map out the receptive fields of recorded cells, and that required the monitoring of eye position. The fact that cats apparently never really fixated on an object appeared puzzling, to say the least. It was bad enough that periods of fixations in primates were filled with a mix between eye drift, ocular tremors, and microsaccades (for a review see Martinez-Conde et al., 2004), but now it seemed that cats didn't even have periods of fixation. How could these creatures see anything at all?

Before it occurred to me that objects could be *tracked*, I thought that the integration of contours faced an insurmountable problem of time pressure. The whole long series of processing steps (as described in the chapters below, including Chapters 19 and 20) had to be finished before the retinal image left the receptive fields of a set of cells. In time, I worked out some of the details of tracking, but could not get away from the problem that tracking failed unless image movement was sufficiently slow.

Then it downed upon me. The cat did not need to fixate its gaze! *The fixations were never meant to make the image "hold still." They were only meant to keep the image movement slow enough that the image could be tracked*!

11.3 A Period of Fixation Is a Period of Tracking

Misnomer or not, the term *fixation* will be used often in this writing, as the events of image organization which take place at the beginning of a period of fixation make up a large part of what I have to say. Saccades (or similar fast eye movements) are accompanied by a surge of activity in many neurons, and some of that activity probably corresponds to what I am describing in the book.

What must be kept in mind is that tracking requires keeping the image drift (at the cortical level, as expressed in millimeters of cortex per second) slow enough that certain processing steps, to be discussed in later sections, are possible. And this means keeping the image movement slow enough everywhere, including the central high-resolution part of the retinotopic projection, where cortical image movement is fastest.

When one realizes this, it becomes clear why cats do not have cleanly defined saccadic eye movements.

The difference between primate eye movement and cat eye movement simply has to do with the fact that the magnification factor is much higher in the central retina of primates (the fovea) than in the central retina of cats (the *area centralis*). In the primate fovea even a very slight movement of the eyes causes the cortical image to race across a large part of the central projection. When the primate slows down its eye movement to the point where the image in foveal cortex can be tracked (Fig. 13.1(b)), it becomes so slow that the eyes appear to hold completely still. The cat does not need to slow down its eye movement quite as much, because the area centralis of the cat speeds up its cortical image movements much less dramatically.

Chapter 12
Theory of the Simple Cell

12.1 Simple Cells, When Detecting LGN Input, Must Link Up *Fast*

A simple cell (ideally) responds only when the contour is centered on its receptive field, not before and not after. It is already noted in the original papers of Hubel and Wiesel (1959) that simple cells are much harder to find in recording sessions than complex cells, because they only respond when the stimulus is exactly in the right place, and they are relatively silent otherwise.

The constant drifting of the retinal image tells us that, in order to keep up with image movement, the contour string must continually recreate itself, and the cells in it must keep being replaced by new ones. As new cells take over, they must first be trained to conduct the contour waves; their synapses must be marked and put in a state of readiness. The need for such preparation has been left out of the illustrations in Figs. 10.1 and 10.3, and the issue will be taken up in this chapter.

The challenge is that the repeated need for the instantaneous creation of new firing chains appears to deny the simple cells the time to be trained, or equivalently denies them the time for receiving the quantity of *surprise* they need in order to mark the synapses which permit them to assume their role in the chain. The simple cells do not have enough time to "warm up" because the instant they detect the lateral geniculate nucleus (LGN) input they must link up with their neighbors and start running the contour waves.

If cells could not join the chain immediately, but had to wait for their upstream input to repeat itself a number of times, they could only join at a rate of one cell at a time, and the chain would grow at a snail's pace. Then each time retinal image drift is called for the replacement of the neurons, each new neuron would need to repeat the procedure.

And it is not easy to arrange a full chain linkup in the instant the LGN input appears, because, as far as is presently known, the LGN input arrives in the form of random bombardment, and entry of a contour into the receptive field merely increases the average rate of bombardment. This in turn means that detecting the contour with certainty, based on LGN input alone, can easily take 100 ms.

It is clear that the network needs to use a trick to get around the dilemma between the linkup being too slow and the surprise being insufficient.

C.R. Legéndy, *Circuits in the Brain*, DOI 10.1007/978-0-387-88849-1_12,

12.2 *Warm-Up* of Simple Cells by the Approaching Contour Wave

The trick as assumed here (Sect. 12.5) amounts to using the contour waves running nearby to tell the cells that the contour will soon get to them, and prepare them to respond immediately when it does. It relies on the fact that, thanks to retinotopic mapping, a drifting contour enters the cortical neighborhood of a cell gradually, and if simple cells have facilitatory contacts with other simple cells with slightly parallel-displaced receptive fields, the facilitation will reduce the chance that the increase in LGN input is accidental (in other words add to the surprise contributed by the LGN input).

Another way to say the same thing is that around the usually plotted receptive field there is a much larger "*subliminal receptive field*," where stimulation does not fire the neuron but contributes to the surprise of its input when the time comes for it to fire. The subliminal contribution is not from LGN cells but from other nearby simple cells which conduct contour waves right before the neuron itself does.

12.3 *Cross-Potentiation*: One Synapse Pool Changing the Effect of Another

The concept that the nearby contour waves should prepare neurons for chain linkup translates into new requirements regarding the individual neurons, going beyond their classical description.

The desired warm-up, when spelled out (Figs. 12.1a–d and 12.2a–f), requires that the synapses on a simple cell be subdivided into "synapse pools," which correspond to functionally distinct presynaptic "neuron pools" (for instance on simple cells the synapses from the LGN are one of the synapse pools), and requires that some synapse pools interact with some others, in the sense that volleys arriving to one synapse pool can change the effect of volleys arriving to another. In the case of interest here (Sect. 12.5), the required interaction causes one synapse pool to exert facilitatory influence on the other one, an effect I will call "cross-potentiation" between the synapse pools. The arrows inside the blobs at the right in Figs. 12.1 and 12.2 indicate *cross-potentiation.*

One complicating issue is that the synapses of different pools are probably not segregated into distinct "synapse colonies" on single dendrites, but are intermingled with other synapses, and are scattered over many dendrites, because synapses form wherever axons happen to cross paths with dendrites from the right cell type.

This means that the synapse pools and the cross-potentiation both require sufficiently specific and free-moving molecular messengers to pinpoint some synapses while ignoring others next to them.

The molecular mechanisms supporting the existence of synapse pools on dendritic trees are expected to be related to the mechanism underlying the *labeled lines* (Sect. 9.1), since both require functionally distinct pools of neurons to be able to pass their chemical signature to the synapses at their output end, and control the nature of their action upon postsynaptic cells. Direct evidence is probably many years away, but evidence does already exist of protein synthesis at nerve terminals, including synthesis in the surrounding glia and (in all likelihood) inside the axons (Alverez et al. 2000; Araque et al. 2001; Piper and Holt, 2004; Genoud et al. 2006; Ge and Duan 2007; Giuditta et al., 2008).

In the case of Figs. 12.1(a) and 12.2(a), input from a series of internally linked synapse pools (from nearby simple cells with parallel-displaced receptive fields) sharpens the effect of input from the LGN afferents, then the resulting enhanced LGN input immediately enables the trigger synapses from the nearby upstream cells (yet-another synapse pool) to fire the neuron (Fig. 12.2(f)).

The sequential arrangement illustrates the way cross-potentiation can augment the *surprise* of a volley not repeated (or, in some cases, not repeated enough times to provide the needed surprise). Instead of being repeated, the volley is preceded by a series of related volleys, arriving in quick succession to a series of internally linked synapse pools which are, when the event is meaningful, expected to receive volleys in the same quick succession before the main volley. When the synapse pools in fact receive their volleys as expected, and the main volley follows, the combined event is a great deal less likely to be accidental than the main volley alone. In this way, the other related synapse pools "pass surprise" to the main synapse pool via cross-potentiation.

The case of the simple cell is one application of the concept of cross-potentiation, but there are also a number others in this book. The need for passing surprise between synapse pools arises in so many different contexts that I feel confident in predicting that cross-potentiation, in the shape of some form of molecular signaling, will be experimentally demonstrated at some time in the future.

12.4 The Graphical Notation of Neuron Sets and Synapse Sets

A few words are in order, describing the "*neuron sets* and *synapse sets* notation" introduced in Fig. 4.2 and used for the first time in Fig. 12.1.

The idea is that each of the amorphous blobs at the right of the diagrams in Fig. 12.1 is a set of neurons (or equivalently the typical member of such a set), and each distinct pool of synapses on the neuron (for instance the synapses from the LGN) is indicated on the blob, near the edge, with a semicircle or rectangle, so that the arrows representing the axons arriving to the synapses can be shown as meeting the synapse pools or synapse sets. In some drawings below, the shape of blobs is used as an indication of the neuron type, for instance half-ellipses (Fig. 19.1) designate groups of contour cells.

Like the neurons themselves, the synapses are drawn as if they were single synapses but are meant to represent the typical member of a synapse set, and the same holds for the arrows representing the axons bringing input to the synapses.

The designs representing the sets of synapses are of two main forms, as in Fig. 4.2; synapse sets already "marked out" on the neuron (probably via long-term potentiation,

Sect. 3.5) are shown as semicircles; synapses not marked out are shown as rectangles. In other words the rectangles represent whole synapse pools which are biologically distinct from other synapse pools, and the semicircles represent subsets of synapses from pools. (In some of the drawings below, where the development of synapse pools is not of interest, all synapse pools and synapse sets are marked by semicircles.)

Blackened semicircles indicate highly reinforced synapse sets, which are such that all volleys arriving to them cause firing, and are sometimes used in supporting synchrony (or co-ignition); semicircles are in some places drawn in thick lines, indicating middle-strong synapse sets causing response at an increased rate.

After a synapse set is marked out and designated with a semicircle, further development of the synapse set is ignored in the drawing (and is mentioned only in comments), so that "successors" of a synapse set (often subsets) are not shown as being different from the "predecessors." In the same vein, synapse sets which gradually change in the course of tracking (as in Fig. 13.1 below) are simply shown as if they were the same synapse set. Tracking is implied in almost all the drawings in Part IV.

The notation in this way indicates when a set of synapses is the successor of another set of synapses, by showing the predecessor and successor set at the same location on the blob, as if they were the same set, whereas in fact they are only *from the same pool*. The biological identity of the pool is what anchors it, and its successors, at a particular place on the blob. The individual synapse choices within the synapse pools, frequently changing under the influence of incoming surprising inputs, are the changeable feature ignored in the notation.

The notational concept of ignoring the changes in set membership is not confined to the synapse sets but is also extended to *neuron sets*. The neuron set represented by a blob changes when a synapse set is marked out on the neurons. In other words, the neuron set on which the synapses are semicircles is smaller than the one on which they are rectangles, because it is made up only of the cells reached by the volleys which do the marking. Here, too, the principle applies that the biological identities of the neurons are constant when the drawings are constant, and the ignored changes are the ones dictated by the ever-fluid surprising events.

The seemingly careless notation, by being blind to the endless progression of subsets within subsets, greatly reduces the confusion inherent in evolving nerve networks.

One feature of the *synapse sets* notation is that it permits us to indicate, using small arrows inside the blobs, the way in which the synapse sets and synapse pools influence each other, as in cross-potentiation. The nature of influence is different in different cases and can change with the operating mode (Sect. 19.3), which means that comments must be added to the figures for clarification.

The rationale behind the loose treatment of set relations is that in each case a biological mechanism is believed to be available for ensuring that the evolving synapse sets are treated as being unchanged when shown unchanged in the diagrams.

As stated, there is no experimental evidence at this point for the existence of the internal arrows (or for the existence of synapse pools for that matter), and, as elsewhere, I proceed on the optimistic assumption that wherever the neurons have a functional need to utilize information available to them in a certain way, they are biologically equipped to do so.

12.5 The Preparation of Simple Cells for Their Role in Contour Waves

Returning to the warm-up of simple cells, it is useful first to make a rough sketch of the steps leading up to the first spike emitted by a simple cell, or at least the steps leading up to the first spike emitted by it as part of the contour wave. This is done in Fig. 12.1; the steps are spelled out in more detail in Fig. 12.2.

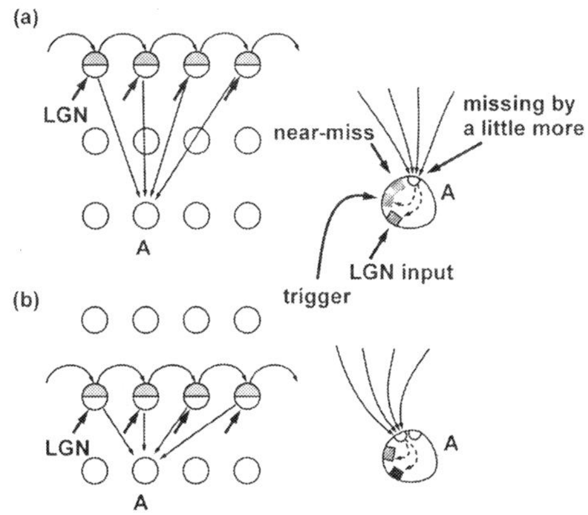

Fig. 12.1(a)–(b) Preparation of a simple cell for contour wave transmission. A contour wave is approaching the test neuron. The drawings show successive versions of an array of simple cells (each identified by a receptive field location) with chains of arrows indicating contour waves passing at different distances from the receptive field of a test cell (A), approaching it. The synapse pools of A and the input arriving to them are shown in insets at the right. In (a) the contour waves are passing within hearing range of the test cell; in (b) the contour waves are passing closer, barely missing the test cell, but do not yet include it. In (a) some of the synapses are already marked through repetition; in (b) more synapses are marked, sending cross-potentiation to the "LGN input" and "trigger" synapse pools

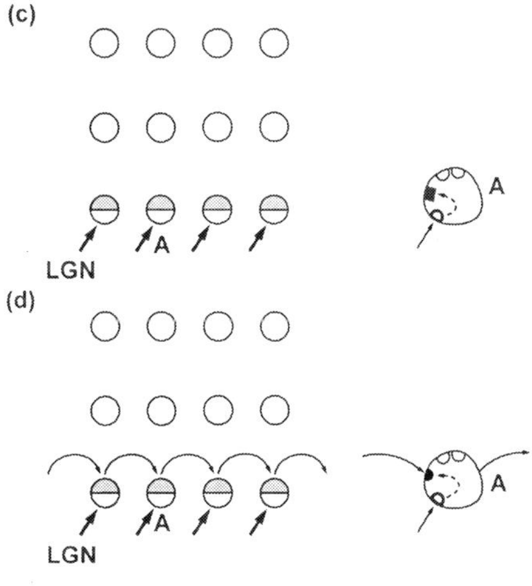

Fig. 12.1(c)–(d) Preparation of a simple cell for contour wave transmission (Cont.). The contour has entered the receptive field of cell. (c) The contour passes through the test cell, as seen from LGN input arriving, but no contour wave was passed through it yet; the test cell is preparing for its role in firing as part of the next event of chain firing; (d) contour wave passes through the test cell and the test cell plays its role in it

The next sequence of drawings, Fig. 12.2, traces essentially the same steps as shown in Fig. 12.1, but in more detail and viewed entirely from the standpoint of the test neuron marked A in Fig. 12.1.

The synapse sets are supplemented by another notational device I find useful, the "input map," which shows the other simple cells which contribute to the synaptic input, by indicating the relative locations of their receptive field centers.

The *trigger synapses* (with asterisks) are so situated that they can act as the links which pass the contour wave (as in Fig. 10.1). To be able to accommodate curvature, they form a slightly wedge-shaped set in the input map, because collinearity imposes a stricter position requirement in the near receptive fields than the farther-away ones.

The input map shows that the neuron is in a position to "know" about its synaptic input, and in this case implies that the neuron "knows" when the contour wave is getting closer.

As in Fig. 12.1, only two sets of "non-collinear" synapses are marked in way of illustration (*near-miss* and *missing by a little more*). The near-miss synapses pass more surprise in cross-potentiation than the ones from farther away (in other words, it is less likely that the arrival of the contour wave right after their excitation is an accident).

The innermost of the not-triggering synapse sets is made up of the opposite-drome in-line synapses (not included in Fig. 12.1 see Fig. 12.2). Their stimulation indicates that the contour is already centered on the receptive field. When these bring input the neuron can be placed into its *wave transmitting state* regardless of detection of the LGN input.

The preparation of a simple cell (A in Fig. 12.1) begins with input from contour waves which pass some distance away but are close enough for their impulses to reach the cell. While only two distinct contour waves are shown here, it can

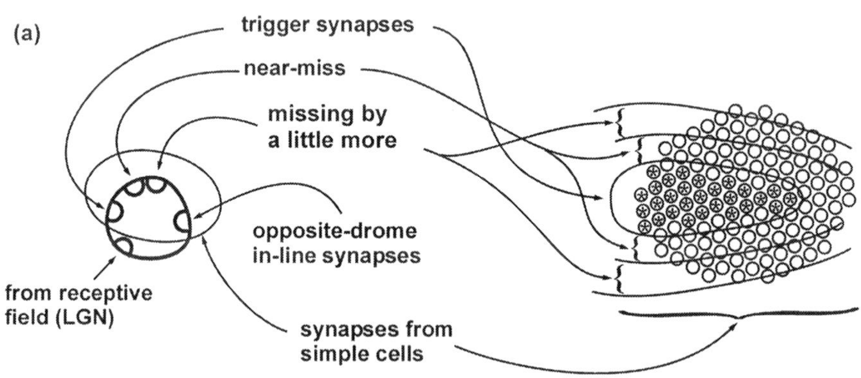

Fig. 12.2(a) The preparation process, as viewed by the test neuron. Notation. Input map and synapse sets of a simple cell. *Left*, "synapse set" drawing of the simple cell. Synapse sets as in Fig. 12.1, with the addition of the opposite-drome in-line synapses. *Right*, "input map" of the simple cell, in which each small circle stands for a synapse on the cell. The placement of the circles indicates the retinal location of the presynaptic cell's receptive field center. The large overlaps between the receptive fields are left out, as are all other receptive field details

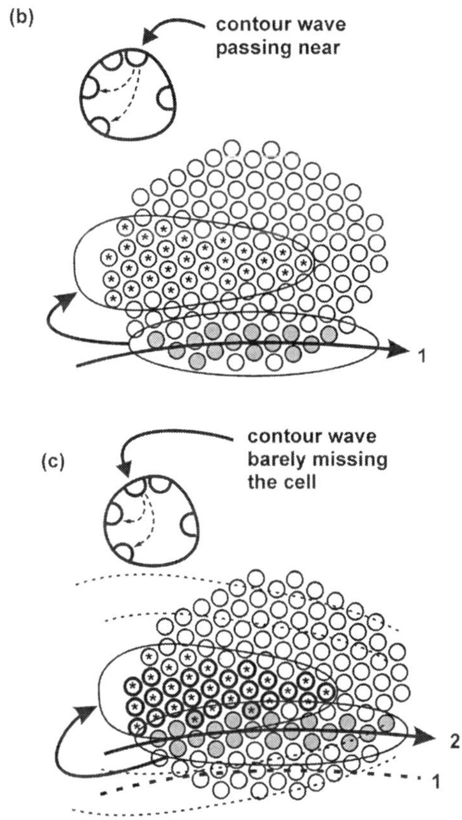

Fig. 12.2(b)–(c) The preparation process, as viewed by the test neuron (Cont.). Contour wave keeps getting closer to the receptive field. Each time a nearby contour wave is detected, through the arrival of a set of spikes in quick succession (from the synapses *colored gray*), the synapses of the trigger set become further facilitated, as are the synapses from the LGN

be assumed that, when the retinal image drift is not too fast, many more pass in-between, and are able to mark out synapses on the test neuron (hence the need for "fixation," which slows down the drift).

The opposite-drome in-line synapses are only excited when the contour is in exact registry with the receptive field, therefore volleys from these innermost members can switch the cell into *wave transmitting state* even before the increased LGN input is detected. It is reasonable to assume, also, that in an input-starved situation, arising at the beginning of a fixation, or when the retinal drift is too fast for clean signals, a near-miss contour wave like the one in (c) can substitute for the LGN input, and send the cell into wave transmitting state.

As a result of the cross-potentiation, the LGN input synapses are made to cause firing more readily. Since the contour has now entered the receptive field and the rate of firing from the LGN is now higher than before, the background firing rate,

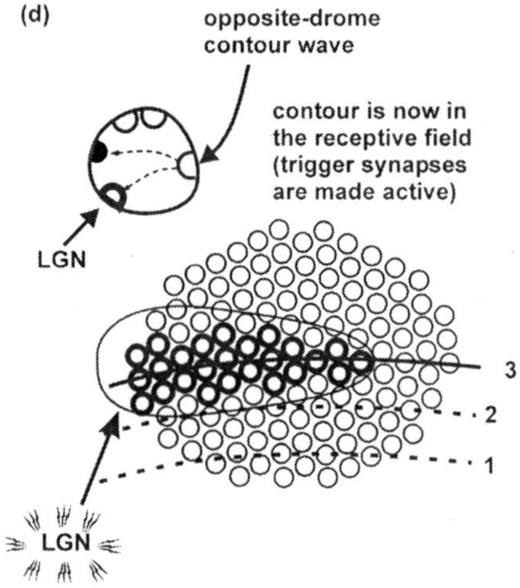

Fig. 12.2(d) The preparation process, as viewed by the test neuron (Cont.). Cell detects LGN input and switches into wave transmitting state. At some point, the receptive field (LGN) stimulation is detected (see also the composite drawing (f)). Because of the cross-potentiation and the recent volleys from neurons whose receptive field is very close to the neuron's own (see (c)), the LGN synapses are now highly effective in their ability to switch the cell into wave transmitting state, where the trigger synapses are enabled to fire the cell

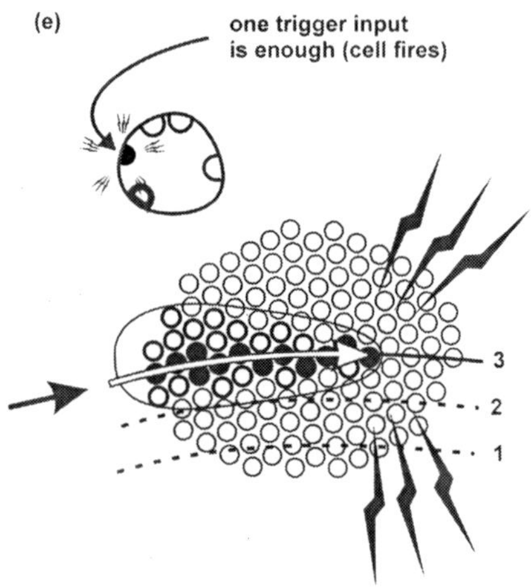

Fig. 12.2(e) The preparation process, as viewed by the test neuron (Cont.). Firing. A contour wave sweeps through the trigger set, bringing inputs through a large number of the trigger set synapses (*white arrow*). Although it is the first time, the trigger set synapses have now such strong weights that as soon as enough neurons of the trigger set (*at right-hand tip*) are reached, the neuron fires

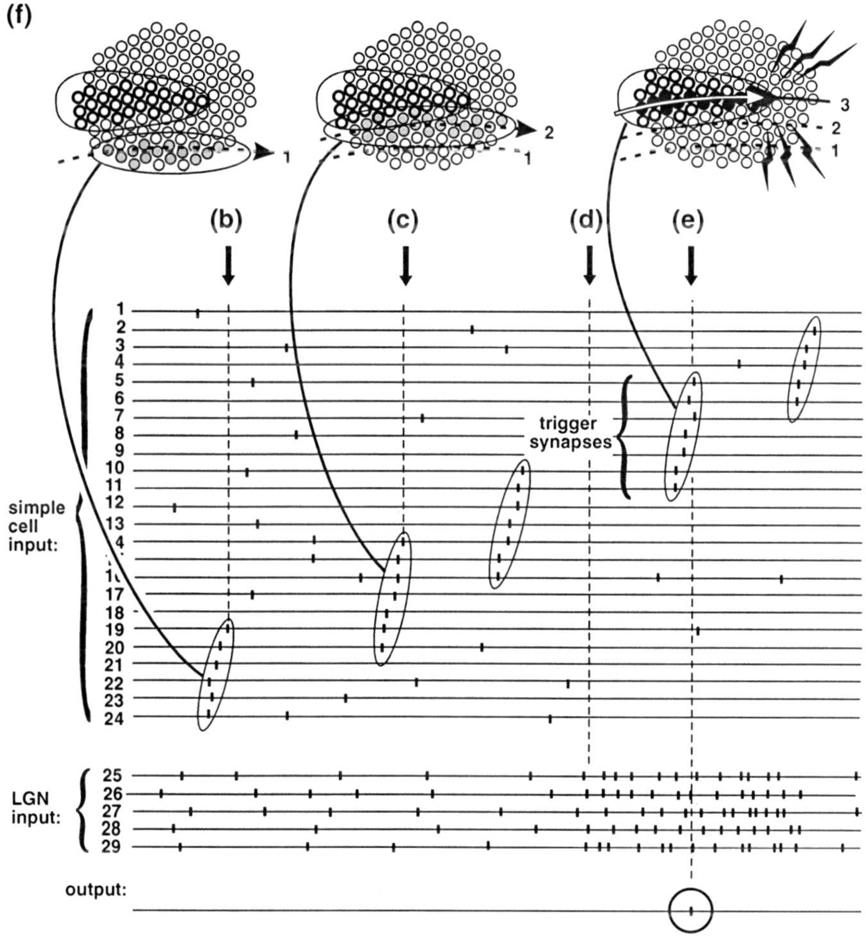

Fig. 12.2(f) The preparation process, as viewed by the test neuron (Cont.). Appearance of spike trains during warm-up and firing. The spike trains show the way the *simple cell input* and *LGN input* to the test neuron may appear during the last phase of warm-up. The input maps from (c) and (e) are reproduced at the top to permit association between sets of synapses circled in the input maps and sets of spikes circled in the spike trains. Arrows on top make reference to drawings (b), (c), (d), (e) above. The *LGN input* spike rate increases at some point, bringing the cell to the "*cell detects LGN input*" phase (see *arrow on top*). The last "run" is accompanied by an output spike. The runs are shown in an "idealized" way, where the simple cell receptive fields are so narrow as to make the contour wave runs almost non-overlapping. In reality, when image drift is not fast, the successive runs are expected to overlap much more than shown; also, the LGN input probably lasts long enough for the cell to emit several spikes

which serves the basis for initiating contour waves (Fig. 10.3(b)), increases. Since at the same time the trigger synapses are made strong enough to propagate waves, the situation shown in Fig. 10.3(b) now applies, together with its two consequences: one, new contour waves are initiated every now and then, and two, contour waves reaching the cell from the left, via its trigger synapses, will cause the cell to fire, taking the contour wave one cell further.

At the very beginning of a period of fixation after a saccade, before contour waves start running, cross-potentiation is not possible because there are no nearby contour waves to facilitate the propagation. Accordingly in a "nascent" contour string the thresholds are expected to be momentarily lowered, and with them the surprise requirement loosened. Once the waves start running, the thresholds can be raised, false starts stopped, and the cross-potentiation applied to its full effect.

The input to simple (as well as complex) cells contains a significant and well-studied inhibitory component (Douglas et al., 1991); however, inhibition does not play a role in the tasks of interest in this book, and is left out of the discussion.

It will be noted that the orderly progression of the spike trains in (f), where each contour wave neatly runs up a staircase, is only achieved through the way the synapses are numbered in the drawing; if the synapses were numbered by their location on the dendrites, the composite would appear "random," corresponding with the way synapses are created on the dendrites wherever the momentary status of dendritic and axonal growth happens to place them. For that reason, it is not obvious that the neuron has a way to know the positions of the cells which go with the synapses, as drawn in the input maps. The underlying assumption is that the successive synapse sets (like the ones circled in (f)) are recorded sometime during early infancy (in some molecular form), for ready retrieval forever after (Sect. 16.5).

Chapter 13
Theory of the Complex Cell

13.1 *Tracking*

The challenges presented by the movement of the retinal image repeatedly bring up the need for *tracking* within the visual system.

The concept of *tracking* is well known to engineers. Earth satellites are tracked by ground stations; aircraft are tracked by other aircraft and by air traffic controllers; tracking appears whenever a changing image must be processed automatically, and processed in a way that tolerates the fact that it is changing.

In the engineering context tracking is a fairly complicated affair, because in addition to holding the image and not losing it, the process also refines the data gathered from the moving object, by means of averaging. The physical laws which govern the motion of the object and the behavior of the measuring system are combined with the latest estimate of the parameters being sought, for instance the components of position and velocity, and used to predict the result of the next measurement. Then comparison with the actual measurement results is used to update the estimate of the underlying parameters.

In aerospace applications the process has been known to yield spectacular levels of accuracy, but it requires high-speed computation and has only become feasible in the age of electronic computers.

In the context of the cerebral cortex the constraints and the possibilities are different; clearly, the nerve net cannot multiply and invert large matrices. Yet, our visual system probably does refine its measurements somehow in the course of tracking, as evidenced by the great visual detail we can perceive. The ways in which the visual system may sharpen its acuity are beyond the scope of the present model; we will only deal with the simplest rudiments of visual tracking. Even those, as will be seen, are sometimes quite involved.

Two parts of the challenge presented by the drifting of the retinal image will be addressed. The first was seen in the example of the simple cells (and will be seen in other examples in Part III); it is the need to "warm up" the cells before being used. The objective is to allow them to gather as much surprise as possible, so that by the time a contour string arrives to their receptive field they are ready to link up and respond immediately, without a need for "rehearsals."

C.R. Legéndy, *Circuits in the Brain*, DOI 10.1007/978-0-387-88849-1_13,

Another and separate part of the challenge is the need for some networks later in the chain of processing to be able to "ignore" the image movement through some technique, and respond only to the properties of the underlying object. It is this second piece where the complex cell becomes necessary, as will be seen.

13.2 Tracking Based on Overlap: *Dynamically Marked* Synapses

The principle on which tracking can be based in the visual context is illustrated in its simplest case in Fig. 13.1a.

Here we consider the input to a neuron that may be created by some set of cells performing an ignition; then a little later a slightly shifted set of cells performing another ignition, then another set, and so on. Each shift is small enough to allow plenty of overlap between the consecutively igniting sets. The tracking consists of detecting the overlap between the sets and responding to a steadily updated set of synapses.

The *dynamic marking* of synapse sets includes a step which shares with long-term depression (LTD) the property that both have the effect of undoing previously created LTP on some synapses. The synapses in the *dynamic marking* situation are first marked after repeated pairing of presynaptic spikes with back-propagating action potentials, then un-marked when the pairing repeatedly fails; in particular, the back-propagating action potentials continue but the time-linked presynaptic stimulation does not.

While the synapses dropped from step to step in Fig. 13.1b are removed through this variant of LTD, the synapses added from step to step benefit from *cross-potentiation* (Sect. 12.4), which helps them to catch up with the rest of the synapse set, by letting them benefit from the surprise transferred from other synapses participating in the previous (overlapping) volleys.

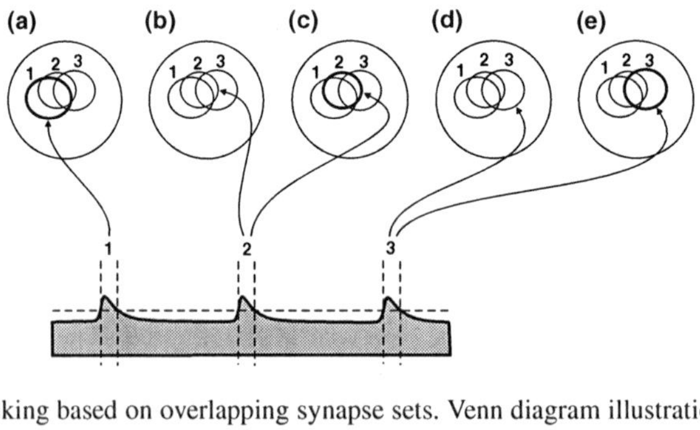

Fig. 13.1a Tracking based on overlapping synapse sets. Venn diagram illustration of tracking by overlap. The sets 1, 2, 3 in the Venn diagrams represent sets of synapses on a neuron tracking a series of ignitions. Ignitions 1, 2, 3 reach the neuron via synapse sets 1, 2, 3, respectively. The same sets are reproduced five times as the system evolves. At each of the ignitions, the membrane potential (schematically shown at the *bottom*) shows a peak (EPSP) acting as a cue to turn the recording on

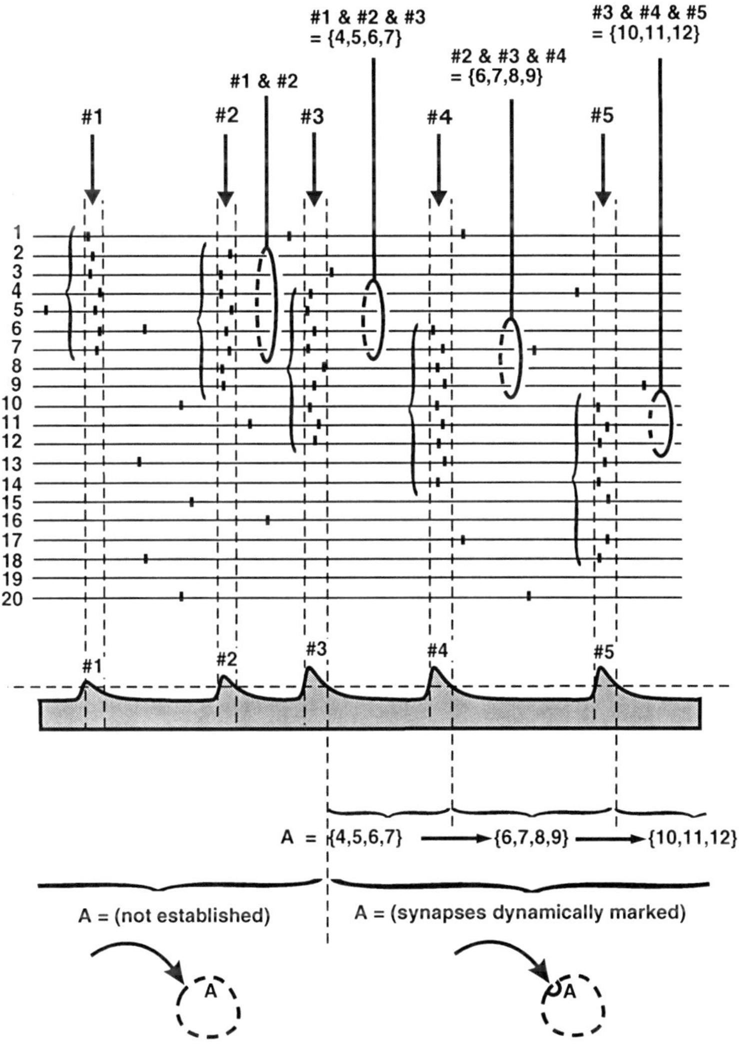

Fig. 13.1b Tracking based on overlapping synapse sets (Cont.). Tracking: confirmation through multiple overlap. Notation as in Fig. 1.6. Twenty spike trains are shown, as they might arrive to synapses of a tracking neuron. At times 1, 2, 3, 4, 5 volleys arrive to the neuron giving rise to EPSPs. The set of synapses repeatedly participating in the volleys would be marked by long-term potentiation (LTP), as in Fig. 1.6, except that there is a difference here: the set gradually changes. Successive volleys have large overlaps, as shown, and there is sufficient surprise to mark sets of synapses, but the marking must be "dynamic," in the sense that it must keep up with the drifting of the set

But, unless I am dead wrong, dynamic marking does a great deal more than just to increase and decrease sets of synaptic weights; it also communicates with the mass storage machinery in and around the neurons with the result that the tracking skills of neurons can improve with practice (see also Sect. 16.5).

13.3 *Dynamic Marking* Shown in Drawings as Just *Marking*

In the "neuron sets and synapse sets" drawings, the "dynamically marked" synapses are shown in a way indistinguishable from the synapse sets ordinarily marked as in Fig. 4.2 and Fig. 13.1; in comments we will refer to the synapses simply as *marked* without emphasizing the dynamic nature of the marking.

And indeed much of the marking in this book is understood to be dynamic. For instance in Fig. 16.1, in the synapse set diagrams at the left of the spike trains, three synapse sets are shown as simply being *marked*, but in fact all three of these synapse sets are *dynamically* marked, as the volleys they receive all come from steadily shifting sets of neurons.

The leisurely treatment of dynamically marked synapse sets is consistent with the purpose of the *neuron sets and synapse sets* notation to remain centered on *functional continuity*.

Dynamic marking may be more frequent in the primary visual cortex than in some other regions of the brain, because of the constant need here to deal with the drifting retinal image.

13.4 The Trick of Simple Cells Feeding into Complex Cells

Returning to simple cells and complex cells, let me mention that the same property of simple cells which enables them to detect continuity also makes their output unsuitable for tracking.

Tracking requires overlap between the sets of neurons which emit successive volleys. Each track in Fig. 13.1b is the output of a neuron, and if no neuron participates in more volleys than one, there is no overlap and no tracking (Fig. 13.2).

At the same time, ideally, the simple cells refuse to respond to a contour element parallel-displaced from their own triggering contour element. In other words, the failure of the (ideal) simple cells to be triggered by parallel-shifted versions of an effective stimulus also means that sets of simple cells responding to two parallel-shifted contours will (often) not overlap, and the tracking will soon fail.

This is where the complex cell enters the picture, as illustrated in Fig. 13.3. The drawing shows a series of contours displaced relative to one another, together with a cell whose input comes from a set of complex cells, some with receptive fields straddling the successive images of the contour. Because of the OR-gate-imitating property of complex cells, their responses to successive contour waves will comfortably overlap.

The cell designated *tracking neuron* in Fig. 13.3 can stand for any of the several cell types; the ones of interest below are the *contour cells*, the *direction-coded cells*, and the *kernel cells*.

As is clear from Fig. 13.1, these cells can only track the contour for a very short time. The objective of tracking at this level of processing is only to hold an image long enough to carry out one phase or another of the operations described in Parts III and IV, and to train new cells to do the rest of the work.

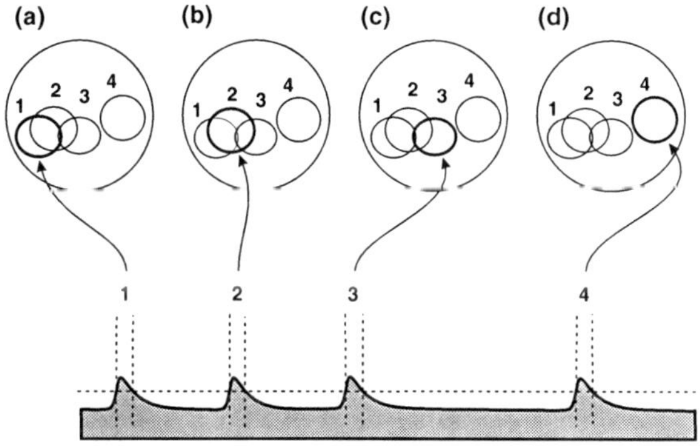

Fig. 13.2 Loss of tracking. It is easy to see that when in a sequence like the one in Fig. 13.1(a) the drifting synapse set moves too far between successive ignitions, there will be no overlap on which to base the tracking (as between sets 3 and 4 here). In the more realistic scheme of Fig. 13.1(b), tracking is lost when the multiple overlap which serves as the basis for tracking contains no synapses, or contains too few to impart the required amount of surprise

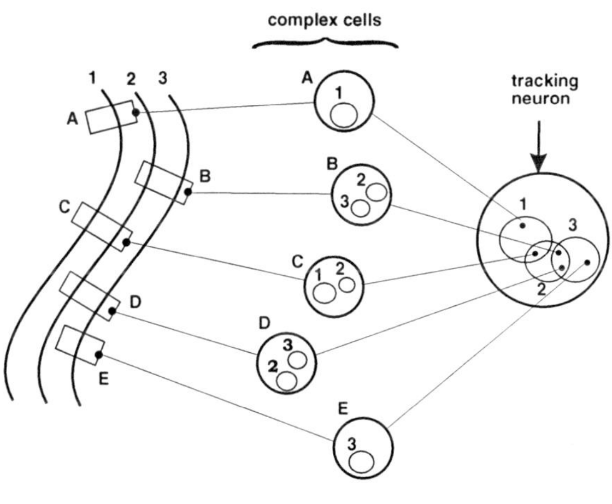

Fig. 13.3 Tracking and the complex cell. A moving contour is shown at the *left* in retinal coordinates; curves 1, 2, 3 are snapshots to the contour at the times when successive contour waves pass. The receptive fields of five complex cells, A, B, C, D, E, are marked in over the contours, showing that some of the receptive fields cover more than one of the snapshots. In the complex cell Venn diagrams, the sets 1, 2, 3 are the sets of synapses invaded by the volleys caused by the three contour waves; in the Venn diagrams of the tracking neuron the sets 1, 2, 3 are the sets of synapses invaded by the complex cell responses to the same volleys. It will be noted that in the complex cells there is no overlap between the three sets but in the tracking neuron there is overlap

When a contour moves slowly enough to be tracked, the snapshots occur much more densely than they appear in this illustration, and accordingly there will be multiple overlaps between volleys from the snapshots, making it possible to confirm the overlaps, as in Fig. 13.1b, and develop a fair amount of confidence in the tracking.

13.5 Simple and Complex Cell Responses Are All *Contour Wave* Responses

The outputs of complex cells are trackable, but the problem remains that the simple cells in contour strings cannot simply be replaced by complex cells, because that would give rise to the failure illustrated in Fig. 10.2. The contour string would be trackable, but would no longer ensure that the cells contributing to the contour waves all lined up on the same contour.

The solution to the dilemma is that complex cells must only respond to the outputs of simple cells *when those outputs occur as part of a contour wave.*

Because the complex cell response is to simple cells participating in contour waves, the collinearity of the simple cell receptive fields remains guaranteed, and at the same time one complex cell can respond to several simple cells with parallel-displaced receptive fields, because they are each in different parallel-displaced contour waves.

According to the present model, during ordinary visual exploration (in unanaesthetized animals), the simple cell response, too, is usually the cell's contribution to contour string activity. The resting firing rate of the simple cell is elevated when the contour is in its receptive field, but, added to the random spikes, there are frequent contour waves, each of them adding a spike to the output of every simple cell in its path, considerably increasing the spike rate, along the same lines as in Fig. 9.12.

Unlike the simple cells, which link up to their downstream neighbors under co-stimulation to perform chain ignitions, the complex cells do not. The only way complex cells can copy the contour adherence property of simple cells is by *passively* following the simple cells as they run each contour wave. It is not permitted for them to be linked up in a separate wave-conducting chain, because chain ignitions propagating over complex cells would fail the test illustrated in Fig. 10.2, and jump from one contour to the other.

It may be suggested that the direct passive response of complex cells to simple cells will often cause them to fire in error, because the complex cells will respond whenever they hear simple cells accidentally firing in quick succession or simultaneously. However, the next layer of cells, listening for chains of complex cell spikes which indicate contour waves, will not be fooled by the erroneous firing of individual complex cells, unless the complex cells before and after them also erroneously fire in contour-wave-like quick succession.

As can seen from the foregoing remarks, the present description of the simple cell–complex cell relation is similar to the classical description of Hubel and Wiesel (1962), but with the twist that in the present model the responses *of both cell types*, at least during active visual exploration, are mostly emitted during participation in contour waves.

13.6 How the Complex Cell Works

It is a known property of complex cells that they typically do not respond to stationary stimulation. This behavior tells us that the complex cell is more than a simple OR gate; otherwise it would respond to any of the constituent rows of simple cells in its receptive field, regardless of image movement.

Instead, it seems, the complex cell contains means to insist on receiving input from more locations than one. This is consistent with the idea raised in Fig. 13.3, which attributes complex cells a role in tracking, and accordingly hinges on each complex cell collecting data from a number of locations of the same contour string.

In the model described in Fig. 13.4, a serial arrangement of cross-potentiation between the synapses from simple cells offers an explanation of the observed behavior, and an arrangement whereby complex cells need inputs from a number of parallel-shifted versions of a contour string in order to fire.

The pattern of cross-potentiation shown in the synapse set diagram of the complex cell describes facilitatory influence between synapse pools from parallel-displaced simple cell rows; in the example of the drawing, where downward movement of the image causes response and upward movement does not, the facilitation is always from sets higher in the figure to sets lower. (In cells responding to image movement in both directions, sets of arrows are envisioned in both directions between the synapse sets.)

The cross-potentiation is arranged in a convergent manner, such that the synapse sets corresponding to sweeps near the bottom of the receptive field receive facilitation from a number of locations above them.

Accordingly, as the contour string enters the receptive field of the cell at the top, as in the present example, its first volleys fail to cause response, since they do not benefit from the cross-potentiation, then the later volleys cause response. This indicates that the thresholds are so set as to require cross-potentiation between *sweeps at different locations*, as a prerequisite for response, even though, from a probabilistic point of view, the same number of repeated sweeps at one location would impart more surprise. This of course only shows that sufficient surprise does not always translate to firing.

In Fig. 13.4 significant overlap is shown between the neuron sets of successive contour waves indicating that here the simple cells are not of the "ideal" kind with zero receptive field width. However, it will be noted that the serial cross-potentiation arrangement accommodates sweeps with very little overlap, as expected for simple cells closer to being "ideal."

Because there is often no (significant) overlap between the successive sweeps in Fig. 13.4, it cannot be said that the complex cell "tracks" the contour wave, and yet the complex cell's very role hinges on its ability to detect that those disjoint sweeps are all part of the same continuum. What the complex cell does is not tracking by overlap, but it is still proper to refer to it as tracking of some kind; I'll call it "*shadow tracking*."

Recalling the remarks on the warm-up of simple cells in Chapter 12, it can be said that the simple cell can also "track" the approaching contour wave (in Fig. 12.2f),

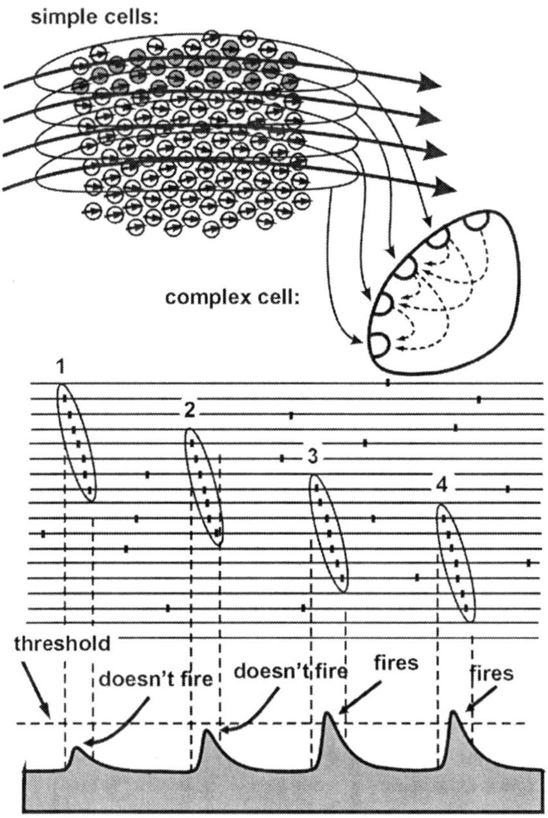

Fig. 13.4 Complex cell: cumulative excitation from sets of simple cells. The input map (*top left*) shows four contour waves (1, 2, 3, 4: the distance between them exaggerated); the cells participating in the topmost one are filled for illustration. The synapse set diagram (*top right*) is drawn large, to show the system of arrows inside, indicating the serial pattern of cross-potentiation. The spike trains (of the simple cells in the input map) show the four contour waves; below them an idealized version of the membrane potential is shown, indicating that the first few sweeps do not cause response

even though there is not enough overlap between the synapse sets invaded by the contour wave. This tracking also falls under the heading of "shadow tracking."

The tracking becomes much easier, and *shadow tracking* becomes unnecessary, if the simple cells have wider receptive fields, and are therefore further from being "ideal." It may be assumed that in early infancy, when the skill of *shadow tracking* is not well developed (see the remarks in Sect. 16.4), and also when image movement is fast, and the warm-up of simple cells becomes difficult, the simple cells have a way to widen their receptive fields, thereby also making both the warm-up of simple cells and the task of the complex cells easier.

Chapter 14
Corner Processing: Theory of the Hypercomplex Cell

The contour wave is the tool for communicating the "figural unity" of a contour to the rest of the brain, and it is reasonable that a sharp turn on the contour should not stop the contour waves dead in their tracks. If it did, triangles and other polygonal shapes could not be broadcast in the way described in Chapter 9, because they would be fractured at each of their corners.

In addition, what appears as a sharp corner when processed in lower-magnification cortex often becomes just another curved section when processed in higher-magnification cortex (for instance the foveal projection). Therefore, if contour waves stop at corners, a shift of gaze can change one contour into two contours. While it is inevitable that magnification should result in changes in the number of objects "seen," it is an important principle of processing that such transitions due to change of magnification be "graceful," meaning that objects must be "trackable" from high-resolution cortex to low-resolution cortex.

It is clearly desirable for the system to enable the contour waves to round the corners of contours; but to do so, certain specialized cell types are needed, other than the simple cells and complex cells. These will be described next.

14.1 Propagation of Contour Waves Toward and Away from Corners

In order to show that it is possible for the neurons to conduct contour waves around a corner, it is necessary to examine the sequence of events in the "warm-up" of the cells at either end of the corner.

Ordinary simple cells in the immediate vicinity of a corner point of a contour only conduct contour waves *toward* the corner point. They conduct (almost) no contour waves *away* from the corner, because for such conduction the trigger cells would have to be (nearly) in the straight-line extension of the line branch of the cells, but the contour makes a turn and leaves all simple cells in the straight extension without stimulation.

C.R. Legéndy, *Circuits in the Brain*, DOI 10.1007/978-0-387-88849-1_14,

14.2 Corner-Supporting Simple (CS Simple) Cells

The ordinary simple cell, in other words, cannot conduct contour waves moving away from a corner point in the immediate vicinity of the corner. To deal with the problem, a variant of the simple cell will be introduced, to be called the "corner-supporting simple cell" (CS simple cell), which does not insist that its trigger synapses come from cells collinear with their receptive fields.

It will be noted that the special cells are only needed for carrying the contour waves *away from* the corners; the contour waves running *toward* the corners are carried by ordinary simple cells.

Unlike ordinary simple cells, which have hard-wired trigger synapses, the CS simple cells must develop their trigger synapses as part of their warm-up. The trigger synapses from beyond the corner are from simple cells; the trigger synapses from the cell-side of the corner are from other CS simple cells which have receptive fields between the cell's own receptive field and the corner.

There are two distinct phases in the warm-up of CS simple cells. The first is the "nascent" phase, soon after a saccade, when contour waves are already running on smooth sections but die at the corners; the second is the "steady-state" phase, when contours are freely running around the corners, and warm-up is possible along lines similar to one described for simple cells.

In the nascent phase, the selection of cells and their trigger synapses makes use of the fact that contour waves are already running toward the corner unimpeded, conducted by simple cells. When at a location the simple cells of one drome conduct waves but their corresponding opposite-drome cells do not, the imbalance is detected by the nearby CS simple cells, whose drome and receptive field matches the nearby simple cells which should be active but are silent.

The same CS simple cells detect, in addition, that the neighborhood contains contour waves of orientation greatly differing from the cells' own preferred orientation; these off-orientation waves within the neighborhood are an indication that there is a corner nearby, or a sharp turn.

The detected imbalance and the off-orientation input, together, enlist these CS simple cells to join the contour wave. After sufficient repetition the enlisted cells mark the synapses from the off-orientation cells as trigger synapses, then start responding to them. As more and more CS simple cells start firing in this way, other CS simple cells downstream from them add their output to their trigger set.

In the steady-state phase, cross-potentiation is possible as before, but the "miss" and "near-miss" synapse sets seen in the ordinary simple cells are not clearly predictable, because of the nature of corners. The retinal drift of an angled contour may sometimes run parallel to the receptive fields on one branch and at an angle to the receptive fields on the other branch; therefore a receptive field on the far side of a corner may lie very close to the contour and yet not be reached by it for a relatively long time.

This does not change the fact that nearby contour waves add to the surprise of the waves arriving to the trigger set. The only difference is that transition to wave-transmitting state is not based on near-miss waves, only on LGN input and the presence of opposite-drome waves. As in the case of simple cells, spurious firing due to erroneous interpretation of inputs will fail to receive confirmation in subsequent wave transmissions and will disappear.

In contour segments having higher-than-usual curvature (as opposed to containing abrupt corners) the CS simple cells and the ordinary simple cells are expected to share in the support of contour waves, because some of the ordinary simple cells will accept the slight orientation difference as being still within their range and some of the CS simple cells will take the slight orientation difference to be large enough to go into active state.

The *CS simple cells* may well make use of feedback from the *hypercomplex cells* (as the simple cells may well utilize feedback from complex cells), but in order to demonstrate that the processing can be accomplished based on peripheral inputs alone, this model does not utilize feedback.

In Figs. 14.1(a)–(f), the *Peer input division* brings input from other CS simple cells. This map is similar to the corresponding input map of the simple cell. *A*, *B*, *C*, are subsets of cells increasingly close to being collinear with the cell of the illustration. The set A brings volleys when a contour wave runs near the cell's receptive field, B does when the volleys barely miss the receptive field (near-miss set). The set C is part of the trigger set, provided that it has any members upstream of the host neuron.

The synapses in the *alignment input division* are from ordinary simple cells whose drome is opposite to the drome of the host neuron. Contour waves of simple cells die at corners, but the drome of these simple cells is such that their waves pass near the cell before they encounter the corner. As a result, these cells generally bring a steady shower of volleys when the contour lies near the receptive field, which makes them suitable for alignment and warm-up. In the nascent phase, when no contour waves come from the direction of the corner, most of the warm-up and cross-potentiation is done through these cells. *D*, *E*, *F*, subsets are comparable to A, B, C, above.

The *side input division* is also from ordinary simple cells. It is not present in the input map of ordinary simple cells, because the preferred orientations of its contributing cells are substantially different from that on the host neuron. Accordingly, a contour wave volley from this division is evidence of a corner (or a sharp turn) nearby. In the *side input division* no receptive field locations have been marked out as being farther or closer to the host neuron, because the host neuron has no built-in way to know where the side input will pass when the contour arrives into exact registry. The only opportunity for gradual improvement is to establish the orientation of the actual side input (*G*) as opposed to noticing simply that some side input exists (*H*).

Since near a corner the ordinary simple cells can only conduct contour waves toward the corner, the side input volleys are always from waves moving toward the corner, and therefore also toward the host neuron's receptive field. This makes these volleys suitable for triggering the host neuron when the time comes. So, as the alignment inputs indicate that the contour is getting closer to its receptive field, the

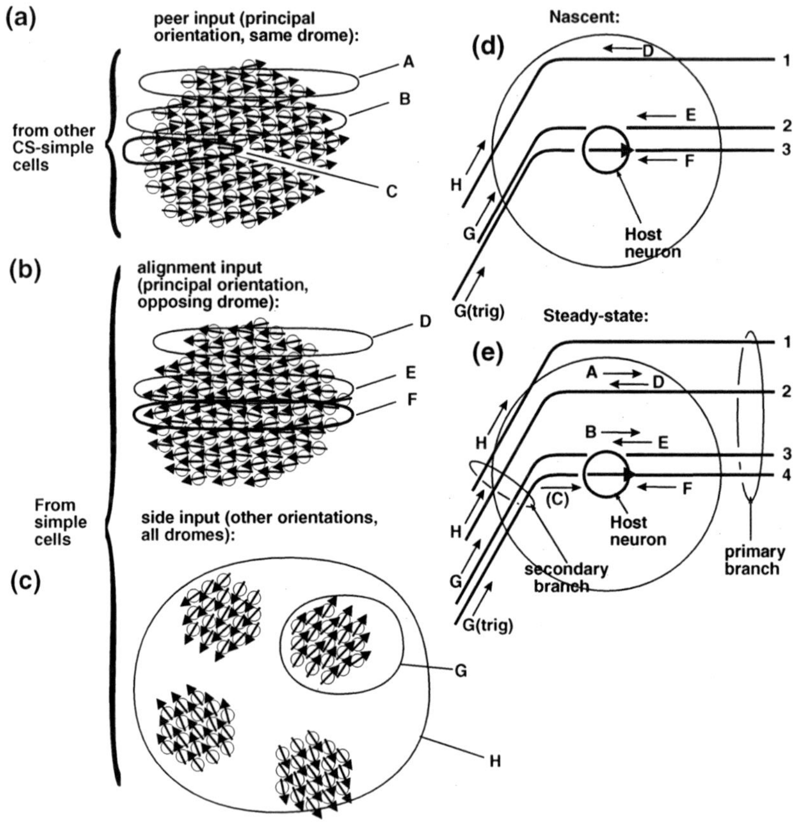

Fig. 14.1(a)–(e) How the CS simple cell works. Divisions of the input map and development of the input sources. At the *left*, the three divisions of the input map are shown; for reference, I will call them the *peer input division*, the *alignment input division*, and the *side input division*. The first two have counterparts in the ordinary simple cell, the third one does not. At the *right*, separately for *nascent* and *steady-state* operation, the drifting bent contour is shown in several stages as it approaches registry with the receptive field of the host neuron (*smaller circle in center, with arrow*). Letters refer to the sections circled in the input map divisions. The *large circle* shows the outer limits of the receptive fields of cells in the divisions. It will be noted that in the host neuron the arrow always points in the direction away from the corner

host neuron triggering mechanism builds up the surprise attached to volleys from the selected orientation of the side input. When the contour reaches registry, the side input becomes trigger input and the host neuron fires.

In Figs. 14.1(g)–(k) and (l)–(p), The first indication (right after a saccade) that a corner is nearby, and that the CS simple cell is likely to be called to action, is that nearby contour waves become one-sided; there are contour waves in one direction (D) and (initially) no contour wave in the opposite direction. The other indication (see (i)) is the presence of nearby contour waves in some radically different direction (H). Accordingly, the synapse sets on H and D are marked. In (h) another

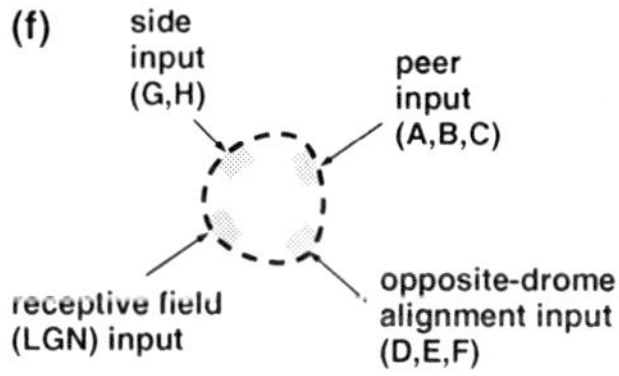

Fig. 14.1(f) How the CS simple cell works (Cont.). Summary of synapse pools in CS simple cells. The gradual build-up of surprise in each of the divisions of the input map, as the neuron warms up and fires, separately in the nascent case and steady-state case, is summarized below in drawings (g)–(k) and (l)–(p), respectively. The capital letters A, B, C, D, E, F, G, H refer to the notations in the input maps (a)–(c) and the contour drift sketches (d)–(e)

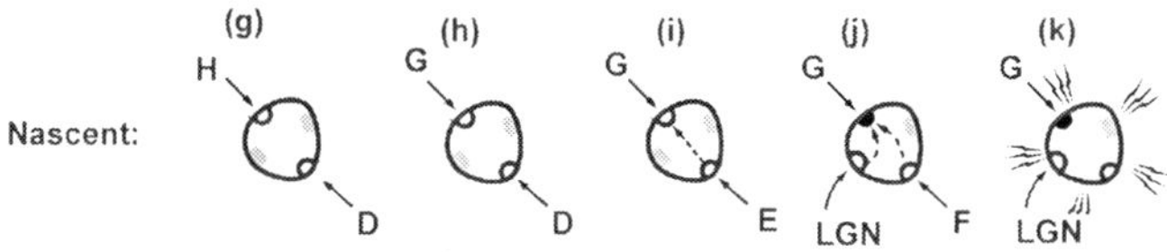

Fig. 14.1(g)–(k) How the CS simple cell works (Cont.). Wave transmission in nascent (startup) operation. (g)–(h) Transition from dormant state to active state; (i) the near-miss synapse set E takes over from D (no separate semicircle is drawn here, to reduce clutter); (j) LGN input is detected; (k) the neuron fires

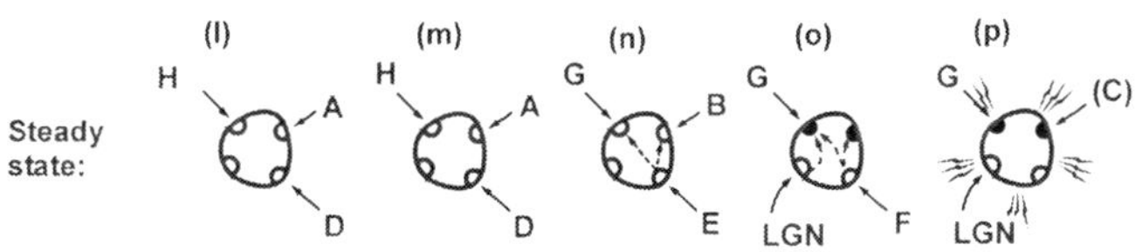

Fig. 14.1(l)–(p) How the CS simple cell works (Cont.). Wave transmission in steady-state operation. The steps parallel those in the nascent operation, but the input sources A, B and possibly C, absent in the nascent case, are now also available. In (p), the triggering and firing step, there is usually some peer input in line with the host neuron and between it and the corner (input (C)); the cells providing this input fire before the host neuron does, and contribute to the triggering

side-input synapse pool, G, corresponding to receptive fields closer to the test neuron's, receives input. It is not designated by a separate semicircle cross-potentiated by H, as the corresponding set is in Figs. 12.1 and 12.2, to avoid clutter.

Similarly, in (i), the near-miss synapse set E takes over from D, indicating that the cell will soon be called upon to fire. The arrow indicating cross-potentiation of the synapse set G indicates that (a set close to) G is about to be made the trigger set.

Then in (j)–(k), the cell makes its transition into *wave-transmitting state*, following the detection that the contour has moved into registry with the host neuron's receptive field. In principle this detection should be based on the receptive field input (LGN input) alone, but since the detection of LGN input can be slow, the volleys from the alignment input division (F) are also used in the detection, and in input-starved situations cause transition without help from the LGN input. Blackening of the synapse set G indicates the highest level of potentiation, in other words that the next input from G (trigger set) is to cause firing. In (k), the next volley from G arrives and the test cell fires.

14.3 Hypercomplex Cells

The CS simple cells share the narrow-receptive-field property with simple cells; they, too, are subject to the requirement to have receptive fields narrow enough not to let them jump from one contour to another nearby contour (Fig. 10.2).

At the same time, the arguments which give rise to complex cells, the need for a trackable output, also gives rise in this case to an analogous wider-receptive-field cell; and this is what is believed to be the hypercomplex cell.

The derivation of hypercomplex cells from CS simple cells is shown in Fig. 14.2. The input map feeds two hypercomplex cells, corresponding to the two branches of the corner. It may be remarked that the case is similar with the complex cells; there, too, two complex cells are derived from similar sets of simple cells, differing in drome only.

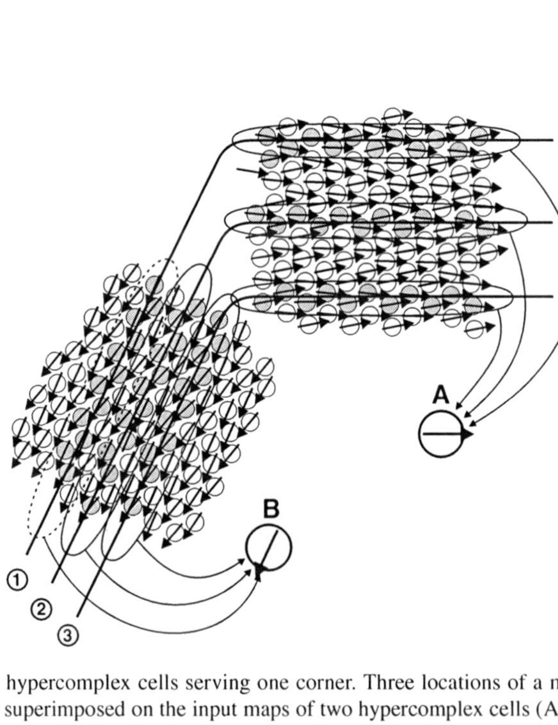

Fig. 14.2 Two hypercomplex cells serving one corner. Three locations of a moving contour are shown (1, 2, 3), superimposed on the input maps of two hypercomplex cells (A, B). The preferred orientation shown in this drawing is always along the principal orientation of the underlying CS simple cells, and the drome is always along their drome (pointing away from the corner)

14.4 Comparing Hypercomplex and CS Simple Cells

It is fair to ask, why the literature (for instance Hubel and Wiesel, 1968) does not contain any reference to a "hyper-simple" cell type.

The answer proposed here is that neither the CS simple cell nor the hypercomplex cell is well suited to receptive field study, because in addition to the primary input, which has a well-defined orientation, each has to have a secondary input, whose preferred orientation is "left open," so as not to narrow down the pool of cells responding to a given combination by too much. Since the CS simple cell has a much wider receptive field than the ordinary simple cell, it is quite easy to believe that both cell types have simply been designated *hypercomplex* in most studies, along with the contour cells and direction-coded cells described in more detail in the next chapters.

Part III
Nodes, Links, Bridgeheads

Chapter 15
Nodes on Contour Strings

15.1 The Problem of Slow Propagation

There is one problem with contour waves: they are slow. Contour waves run from cell to cell over very short axonal links, and these are generally not myelinated. Myelinated axons can conduct spikes at a rate of several m/s or more, but on non-myelinated axons the speed of propagation is only about one tenth of that.

As a result, it can take as long as 50–100 ms for a contour wave to propagate from one end of the contour string to the other, too long to support the creation of single ignition events combining information from far-apart points of a contour.

The slow propagation of contour waves eliminates the simplest way to communicate the co-belonging of a contour string to the rest of the cortex, the sudden and crisp ignition by the entire chain. If the neurons cannot fire in quick succession, their ignition will at best be a slowly developing event which cannot stand out against the random background.

The need arises for a fast way to interconnect the different parts of a contour string.

15.2 The Stria of Gennari

Now, it so happens that there is no shortage of fast myelinated axons running horizontally in the primary visual cortex.

They form what is the most noticeable of the intracortical fiber systems within V1, the stria of Gennari (Gennari, 1782; Vic-D'Azyr, 1786; von Bonin, 1960; for a review of the literature see Schmidt and Löwel, 2002). It is a massive band of myelinated fibers classically described as running horizontally in the upper portion of layer 4, which is visible to the naked eye, and is the reason V1 (or area 17) is also called the "striate cortex." The recent literature (Stettler et al., 2002; Binzegger et al., 2004) tends to avoid reference to the stria of Gennari when writing about the long horizontal axons in the upper part of the primary visual cortex, maybe because the old designation is too restrictive, but in the present writing I am using the old designation, because of its convenience, with the understanding that the classical descriptions can be updated where necessary.

C.R. Legéndy, *Circuits in the Brain*, DOI 10.1007/978-0-387-88849-1_15,

Ramón y Cajal (1911) had believed that the stria was made up of geniculate afferents, but Szentágothai (1973), in an experiment where he undercut the cortex and examined the degeneration patterns, showed that the origin of most fibers in the stria was intracortical.

It is now known that the stria (as classically defined) is largely made up of axon collaterals from the pyramidal cells of layer 2/3 and the upper part of layer 4 of cortex (see also Stettler et al., 2002). These cells send their apical dendrites to layer 1, and their axons to the white matter, with long horizontal collaterals (Creutzfeldt et al., 1977) joining the stria. The fibers of the stria are believed to terminate on the same anatomical group of cells as are their sources (Shkolnik-Yarros, 1961; Valverde, 1971).

In addition to being connected through the intracortical fibers of the stria of Gennari, upper layer visual cortical cells are also believed to be connected through "U-fibers," which are fibers descending into white matter and re-emerging elsewhere within the same cortical area. Superficial lesions in the visual cortex, affecting layers 3 and 4, have been found to result in degeneration of numerous U-fibers (Szentágothai, 1973), and U-fibers have been described as originating as well as terminating on layer 3 cells Shkolnik-Yarros, 1961). Apparently, when the source cell and the target cells are more than a few millimeters apart, the axon does not join the stria of Gennari but goes down into the white matter and re-emerges near its target cells. Although not part of the stria of Gennari in the literal sense, these axons appear to belong with it from a functional point of view, as they are also part of the system of long fast fibers linking upper layer neurons with each other.

Accordingly, in this writing I am using the term "Gennari fibers" loosely, to designate all fast horizontal fibers between layer 2/3 pyramidal cells, including the U-fibers.

15.3 Speed-Up by Means of *Nodes* Linked by Gennari Fibers

The next question is how can the fast horizontal fibers help in creating ignitions which connect different parts of a contour. In this model the question is addressed through a two-tier arrangement built around the contour string, where cells first form "nodes" on the contour, made up of cells close enough together to ignite fast, and then pairs of nodes on a contour link up to one another through the fast Gennari fibers.

The advantage of nodes is that by coordinating their ignitions they can communicate more than just the fact that they are on the same contour; they can also communicate shape information as was illustrated in Chapter 9.

The neurons making up the nodes are not the same as the simple cells and the complex cells; rather, as will be seen, they form a separate and dependent system which only utilizes the complex (and hypercomplex) cell input for information about the presence of a contour nearby.

15.4 Nodes Viewed as Representing *Points*

The nodes on a contour string are meant to represent *points* on the contour. The cells of a node are all selected from one column (CO blob-interblob system) or, if one column does not have enough suitable neurons, from a number of adjacent columns. It will be recalled that the width of the cortical column corresponds (by and large) to the smallest perceptually resolvable distance, which makes it suitable as the network representation of a point.

One may ask, if a node is just a point, a structureless thing of zero extension, why would it need a whole column's worth of neurons, and how can that many neurons even be insufficient?

The answer is that in visual processing the information contained in a point is, to a large extent, information dealing with the point's relation to other points. The point itself is of not much use in (for instance) a directional relation without the other point and without the fibers reaching from the network of one point to the network of the other point.

But if the location of a retinal point and the location of its corresponding node are linked through retinotopy, it follows that the retinal image of a contour dictates the cortical distance between its nodes.

This in turn raises the possibility that two nodes, especially in primates, will be several centimeters apart in the cortex, and the number of axons originating in one and synapsing in the other will be insufficient to support co-ignitions.

In other words, far from having too many neurons, the cortical column, if anything, has too few.

15.5 A Note on the Fiber Requirement of Visual Integration

It is reasonable at this point that we take a look at the anatomical data and estimate whether a cortical column in fact contains enough neurons to reach all other columns in the primary visual cortex (and find that it often doesn't).

The available data tell us that in the macaque there are roughly 5,000 blob/interblob sets per hemisphere (Purves and LaMantia, 1993).

There are about 250,000 neurons per square millimeter in the visual cortex (Blinkov and Glezer, 1968), and about 5 blobs per square millimeter (Purves and LaMantia, 1993). If the cells are divided between blob and interblob regions in a 20–80% ratio, and only the interblob regions contain the cells of interest, we have 40,000 neurons per column. The cells of interest are presumably the layer 2/3 pyramid cells, which have been described the most numerous cell pool in the cortex (Binzegger et al., 2004). Only a subset of them can be assumed to be direction-coded cells with long horizontal axons, say 10,000 at the most, to distribute their axons among all columns. If every axon makes "clustered" branching in 5 columns, the average column, very roughly, sends 10 axons to the average other column.

The estimate, while crude, is good enough to tell us that while (in the primate brain) the average column may well reach most columns in its immediate neighborhood, it is unlikely to reach most columns farther away, especially when we consider that reaching a neuron probably requires concentrating enough synapses on some dendrite of the neuron to start dendritic spikes in it (Losonczy and Magee, 2006; Remy and Spruston, 2007).

As a result we must envision the typical node as being made up of a number of columns.

In the mouse (Braitenberg and Schüz, 1998), where the brain is much smaller, the columns can (possibly) be expected to reach all other columns, so the nodes are probably made up of single columns; however, in primates the same cannot be assumed.

Considerations such as these will serve to illustrate the enormous fiber requirement of visual image integration. In addition to the need for nodes to reach other nodes, the fiber requirement in the first stage of visual integration, the stage described in this book, is increased by the need to keep up with the constant drifting of the retinal image. (It may be assumed that starting with V2 the information is deposited in places which, while distributed and redundant, do not move around.)

If there are too few available axons in V1, only a fraction of the node pairs is expected to succeed in linking up, and accordingly the linkage is expected to depend on the momentary set of cells in the node pairs. Since image drift will soon move the nodes away from their optimal location, their linkage is likely to be lost before the nodes can develop co-ignitions and broadcast their directional relations.

Most cortical areas contain long horizontal systems of myelinated fibers in their interior, the striae of Baillarger (1840). In the sensory cortex the outer stria, which in V1 is the stria of Gennari, deals by and large with the integration of incoming sensory information; the inner stria, by and large, with the integration of information sent out centrifugally or toward the motor division. None of the striae of Baillarger is as visibly massive as the stria of Gennari.

15.6 The Placement of Nodes on a Contour

It is not easy to guess what strategy the brain uses when selecting the location of nodes on a contour.

Better said, one part of the answer is easy to give: when the contour has corners, the corners should be among points chosen for nodes. So should be the tips of high-curvature segments. The corner-supporting simple cells and the hypercomplex cells pick out these locations quite reliably, and put them at the same place on the contour over a wide range of magnifications.

The more difficult part is to assign nodes to the gently curved segments, or to the straight segments, of contours, because node assignment strategies must be well enough anchored in the contour shape that each appearance of the same contour on the retina will lead to nodes making similar polygons.

Part of the problem is that any fixed scheme of node selection imposes a limitation on the accuracy to which a shape can be described. The limitation can only be eliminated by centrifugal interrogation of the cortical image from higher cortex, and such interrogation is almost certainly present in real brains. But centrifugal signals are left out of the present simple model.

As it happens, the difficulty in deciding on a good method of node selection is not a "show stopper," since in this book our interest is centered on the things done with the nodes once they are selected.

The working hypothesis used below is that nodes on smooth sections are selected at columns where the contour waves match one or another set of preselected slopes, chosen to correspond to the slopes of the sides of a regular polygon.

Refinements upon the node assignment using feedback from upstairs, which are almost certainly central to the process, are left to future work.

15.7 A Link Between Nodes Has a *Bridgehead* on Each Node

Continuing with the properties of "polygon graphics," and the requirement to join a number of sentences on one noun (one node), we note that in a drawing of nodes with diagonals crisscrossing between them, as in Fig. 9.7, each node is expected to have at least two lines converging on it, and many will have more than two (Fig. 9.8).

This means that, in the neuronal representation, one node must be able to host many links. In fact, ideally there should be no built-in limitation to the number of links a node can make; the design is expected to make it possible to keep adding links to a node as the need arises.

A link between two nodes is made up of a number of axons between the nodes; it is a physical pathway comparable to a *bridge* connecting the nodes. The bridge can be traversed in either direction and has a "bridgehead" on each of the two nodes it connects.

In other words, a node must contain at least two kinds of neurons: the bridgehead cells with long intracortical axons (contour cells or direction-coded cells), and the kernel cells with short intracortical axons, which are shared between all bridgeheads of the node.

The building block of the bridgeheads is a long-axon neuron (by which from now on I mean long *intracortical* axon or U-fiber) with its cell body located in the node hosting the bridgehead, and its axon, which is part of the stria of Gennari in the extended sense, making synapses on the opposite bridgehead. A *link* is made up of two bridgeheads, one in each node, with mutual synaptic contact. The two nodes are connected by enough fibers that either bridgehead can ignite the other.

The arrangement has the fortunate feature that the long axons of different bridgeheads tend to arborize in different cortical neighborhoods, and therefore even when a node has quite a few bridgeheads they are (often) made up of non-overlapping sets of cells. Therefore there is very little interference between the bridgeheads in a node and their ignitions do not disturb each other. This is true, by and large, even when a node happens to be linked to two nodes within the same sector of directional angles (Fig. 8.1), because direction-coded cells, even when "clustered" (Gilbert and Wiesel, 1983), do not cover every column in their sector equally well.

Incidentally, the functional need for bidirectional fast-fiber linkage between direction-coded cells has the corollary that the bridgehead cells need to be able to act both as sources and targets of the fast horizontal fibers. The existence of a large pool of cells satisfying this requirement is supported by the anatomical and physiological findings just mentioned (Shkolnik-Yarros, 1961; Valverde, 1971; Gilbert and Wiesel, 1979; Ts'o et al., 1986) which indicate that fibers of the stria of Gennari both originate and terminate in cells of layer 2/3, where the direction-coded cells and contour cells are believed to exist.

With this in mind, one more piece of the puzzle falls into place. It is known that the axons leaving the primary visual cortex for visual area V2 (Brodmann's area 18) mainly originate layer 2/3 (Gilbert and Kelly, 1975). If the cells making up the bridgeheads are mainly layer 2/3 pyramids, they are in the proper position to send their visual sentences to the cortex upstairs.

Chapter 16
Custom-Made Unstable Networks Made to Support Tracking

Returning for a moment to the subject of tracking, and recalling the role of complex cells as devices to make the outputs of simple cells trackable, it will be noted that the smooth movement of retinal images does not in itself guarantee that the images can be tracked.

It may be tempting to assume that retinotopic mapping renders the task of tracking trivial, because the moving images give rise, in essence, to moving "clouds" of neuronal response which are bound to overlap between successive snapshots.

This would in fact be correct if the neurons of the visual cortex did nothing more than passively follow the inputs coming from the lateral geniculate nucleus (LGN), as in earlier models of vision.

But the situation changes when one realizes that the primary visual cortex needs to contain "prime movers," which are essentially custom-made unstable networks whose role is to generate the surprising multi-neuronal signals which the other parts of the network can detect.

The contour string is one example of such an unstable network; it is able to generate surprising firing in the form of waves of spikes running down along a line (Fig. 10.3b), all because of the (assumed) way in which simple cells act when co-illuminated.

As was seen in connection with the interplay between simple cells and complex cells, the active event-generating networks do not automatically offer overlap between the volleys they generate at successive positions of the contour; in fact the outputs in general cannot be tracked unless they are first "massaged" in some way (Fig. 13.3).

16.1 Self-Igniting Networks Which Continually Gain and Lose Cells

It can be said that, as the retinal image evolves, a sequence of hardware devices must be created and recreated, and at all times these devices must satisfy the requirement that the coordinated events they generate are *trackable*.

C.R. Legéndy, *Circuits in the Brain*, DOI 10.1007/978-0-387-88849-1_16,

The term "*maintenance of neuron groups*" will be used for the continuous operation whereby neuron groups cope with the drifting retinal input. Part of the role of maintenance is to see to it that the outputs do not change drastically from one multi-cell event to the next, and are therefore trackable.

In the context of nodes, and the various ignitable groups in them, the maintenance work implies continually adding new neurons to the groups and dropping old ones, while updating the synaptic contacts between them so that the newly added neurons join the ignitions of the old neurons, and thereby make their co-belonging noticeable.

16.2 Active Linkage: Two Bridgeheads Repeatedly Co-igniting

Nodes and the linkages between them must be visible to the rest of the network. They cannot for instance just sit there quietly, and wait to be interrogated in some way before announcing themselves and their mutual connection. The constant movement of the contour does not allow enough time for such interrogation, especially since the cells receiving the interrogation would first have to mark the synapses on which to receive it.

Nodes cannot be envisioned as being permanent objects. They are more along the lines of the "endless poker games" which, we are told, existed in the Wild West, where new people kept joining the game while others kept dropping out. The only thing constant was the poker game itself.

In the neuronal context this requires a fair amount of processing, as will be seen in more detail in Fig. 18.1. The selection and training of new neurons for their roles in the nodes must proceed quite fast, which means that neurons near the nodes, ones which may have to take over if the contour moves into their receptive field, must be prepared for the roles in advance with the proper synapses marked, even though it is not sure that they will ever be called upon to use the training. In addition, the nodes and the links connecting them must change gracefully with plenty of overlap between their successive versions, so other remote structures can track them.

As was mentioned, it is desirable that the nodes and the link between them repeatedly and continually keep announcing their latest membership, which is just the kind of action for which the repeated *self-ignitions* (Fig. 7.1) are ideal.

The need for continual "visibility" of the links introduces the concept of the "*active* link," a link made up of two mutually connected bridgeheads, each in a different node, each performing repeated random self-ignitions, and each in addition set up to follow, passively, the other bridgehead's self-ignitions. Roughly one-half of the ignitions are initiated by one node and half by the other.

Two-way active linkage, in which each node passively follows the self-ignitions of the other, is only feasible because the self-ignitions run on random schedules. If self-ignitions were periodic, the autonomy of the sides would be lost. One may note in this connection that the self-ignitions at the two sides are not *independent* random processes, because whenever one group passively follows an ignition from the other group, its "timer" is reset, schematically speaking, and the responding group will only be able to initiate (or follow) another ignition after passage of the required "dead time" (refractory period).

This means that quite often one side ignites and the other side fails to follow. It is important to note that by the nature of surprise-based communication such failure is tolerable from the standpoint of information transfer, even if it occurs frequently, because the surprise is solely determined by the frequency of *successes*. Failures are no different from any other noise and as long as the rate of successful co-ignitions sufficiently exceeds the expected rate, they will carry the surprise needed to convey connectedness.

It may appear that the continually repeating co-ignitions of a node would interfere with one another, but it must be noted that the different bridgeheads send axons to different places, and are accordingly generally hosted on non-overlapping sets of cells; since they run on separate tracks, their self-ignitions do not interfere.

The two-sided "hammering" in an active link makes it possible to track the nodes, train new neurons, and shed unnecessary columns when a bridgehead becomes too large (Fig. 17.1).

This suggests that the contours whose links are maintained as *active links*, beyond the time necessary to support general linkup, are the ones on which *visual attention* is momentarily directed. However, the discussion of visual attention requires introducing the "centrifugal" influences acting on the primary visual cortex, in other words the inputs coming from higher cortex; therefore the subject is left out of this model.

16.3 Detecting When a Link Becomes Weak

In order to maintain a link in the face of ever-changing node membership, it is, before anything else, necessary for the elements which are to correct a problem to be able to find out that there is a problem.

It is desirable to be able to detect a problem with a link before the linkage is totally lost; in other words when the link is merely weaker than it should be. In the next sequence (Fig. 16.1a) it is shown that, in active links, thanks to the repeated co-ignitions, the typical neuron of each bridgehead can (in principle) separately monitor its incoming path and its outgoing path, and detect when either of them becomes weak, or both do.

It will be noted that in the present context the problem is considered solved as soon as it is shown (as in Figs. 16.1(b, c, d and e)) that the needed information is available at the locality where corrective measures are to be initiated (Figs. 16.2 (a, b, c and d)). In fact, of course, showing this does not solve the problem at all, only experimental evidence solves it, describing the details of the detection and the corrective measures. The present "solution" (along with similar ones in Figs. 19.3, 19.5, 19.6 and others) is merely meant to demonstrate that the processing is geometrically possible.

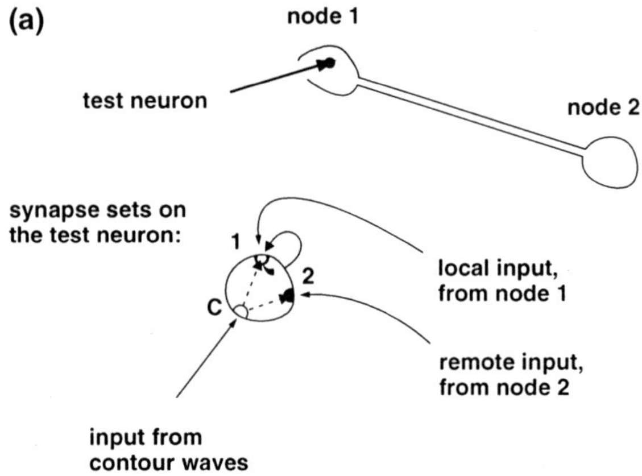

Fig. 16.1(a) Weakening of an active link, as viewed by one neuron. Nodes 1 and 2 are actively linked; the test neuron is in node 1. The next few drawings deal with the form of the spike trains arriving to a neuron in one of two interconnected nodes, under various conditions, and demonstrate that when the link between the nodes becomes weak in either direction, the neuron has a way of knowing about it. *Bottom*, synapse sets on the test neuron (all dynamically marked), reproduced in each of (b)–(e)

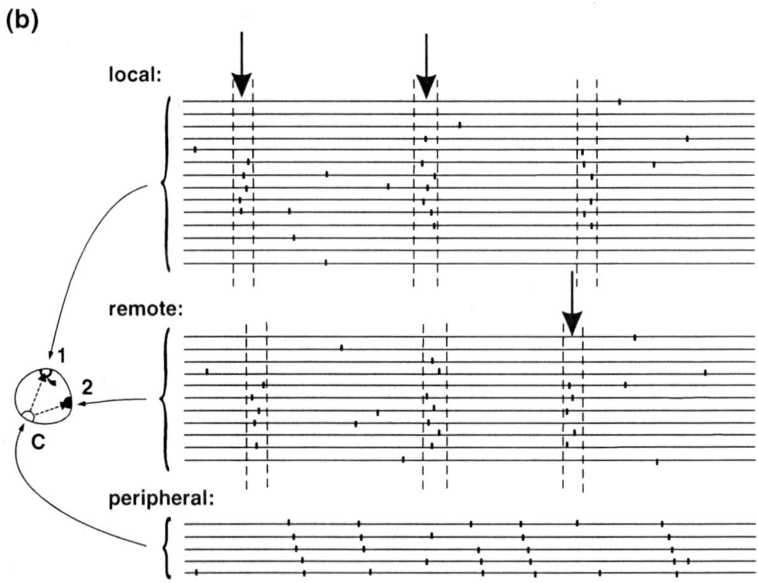

Fig. 16.1(b) Weakening of an active link, as viewed by one neuron (Cont.). Link between nodes 1 and 2 is strong in both directions. Three sets of spike trains are shown as they arrive to the three synapse sets, 1, 2, and C (see inset). The *arrows at the top* show where the self-ignitions are initiated; when they are locally initiated (in node 1), they are shown on the top set of spike trains; when they are remotely initiated (in node 2), the *arrow points to the middle* set of spike trains. The passive response is slightly delayed, because of conduction delay in the Gennari fibers, so in each pair of volleys the volley marked by *arrow on top* is slightly ahead of the other member of the pair. Because of the bidirectional link, volleys initiated in one node are passively followed by the other; therefore, when the link is strong in both directions, each volley initiated in one end of the link is soon followed by a volley from the other.

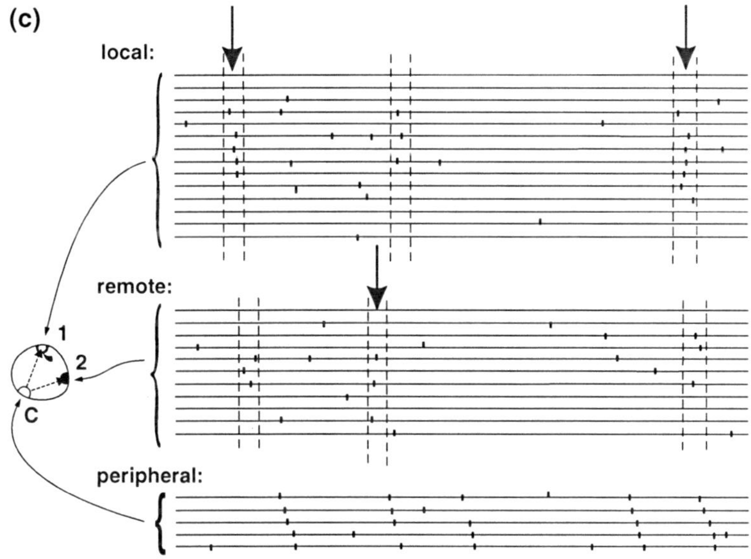

Fig. 16.1(c) Weakening of an active link, as viewed by one neuron (Cont.). Link 2→1 is getting weak. Since the link from 1 to 2 is still strong, the self-ignition on the top traces reaches node 2 and causes a strong ignition in node 2. But the echo to these is weak, because the link from 2 to 1 is weak, and many of the cells in 1 which are supposed to follow the ignition of 2 fail to do so, therefore a weaker response is seen in the second set of tracks (broken lines indicate the time intervals where the responses should be). Similarly, when the self-ignition is generated in node 2, it is strong, but, because of the weak 2→1 link, it is not heard well in node 1. The result is that the second set of tracks shows an incomplete set of spikes

The "recurrent collaterals" of the axons (shown, for instance, in Fig. 1.11 (d)) support the ignitions of the bridgeheads (self-directed arrow at synapse set 1). The selection of axons reaching the opposite node is accomplished previously during the linkup (Figs. 19.4 and 20.2).

Each bridgehead repeatedly executes self-ignitions on a random schedule ("crescendo" flag on synapse set 1), and each bridgehead receives strong enough input from the opposite bridgehead that it passively follows its ignitions (blackening of synapse set 2); so each self-ignition results in a co-ignition of the two bridgeheads.

It is assumed throughout this model that the local input to the neurons is chemically distinguishable from the remote input (as in the drawings (b)–(e) below). It is not obvious that this is possible, because it means that a cell is able to send out one kind of chemical identifier over its synapses near the cell body and another kind over its synapses far.

The reason for the somewhat unconventional assumption is that the model relies on the ability of bridgeheads to make autonomous *local* ignitions, and also on their ability, as in the present series of drawings, to treat local ignitions differently from the ignitions occurring at remote nodes.

While it is possible to think up ways to arrange autonomous local ignitions with the help of interacting pools of short-axon neurons, such solutions would be slow

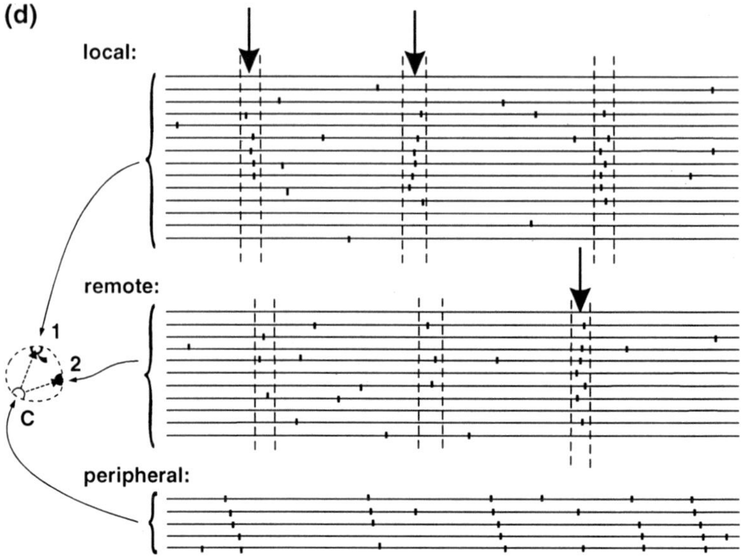

Fig. 16.1(d) Weakening of an active link, as viewed by one neuron (Cont.). Link 1→2 is getting weak. The situation is distinguishable from (c) by the fact that only the echo of the locally initiated volleys is diminished. The remotely initiated volley comes through undiminished, because it sends a full volley to the local node and the local node makes a full response to it

and clumsy, which Nature's solutions seldom are. I predict that future research will identify a mechanism for implementing the near-synapse–far-synapse distinction.

The next few drawings show the spike trains received by a typical bridgehead cell with its cell body located in node 1. The synapse set marked "1" is from axons of local origin (node 1), the set marked "2" is from axons coming from node 2, and the set marked "C," the peripheral input, is from the axons of the nearby complex (or hypercomplex) cells bringing occasional runs of spikes in quick succession from contour waves, and thereby letting the test neuron know when the contour appears beneath it, and when it is gone.

The test can be considered an "echo test," because when the arrow is over a volley in the top set of traces, the corresponding volley in the middle set of traces is an "echo" received by the test neuron to the locally generated volley.

In the case of locally generated ignitions, the test neuron participates in the crescendo action during build-up to self-ignition, and accordingly "knows" that the ignitions are locally generated. When the subsequent remote ignition arrives to its synapses, which are set up for passive following, the cell does not respond a second time because of refractoriness. In the case of the remotely generated ignitions, where the test neuron does fire in passive response, once again the test neuron "knows" that the volley is remotely initiated.

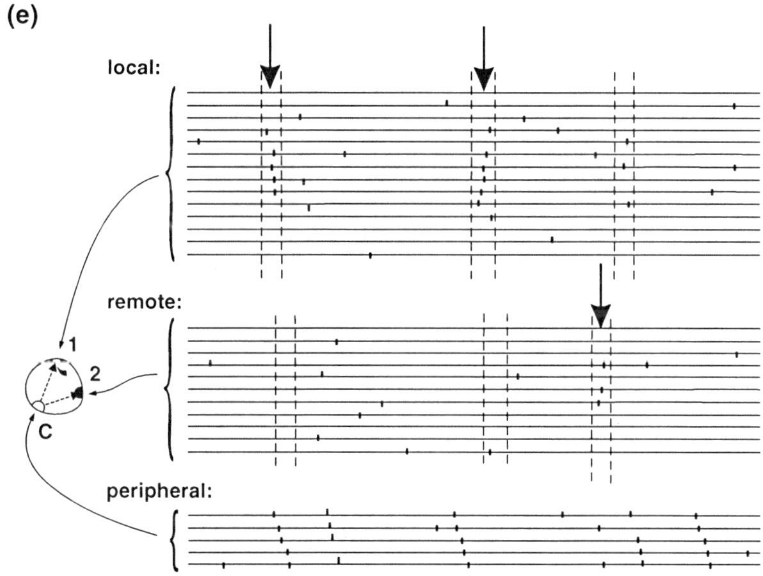

Fig. 16.1(e) Weakening of an active link, as viewed by one neuron (Cont.). Links weak in both directions. In this case the echoes to the locally generated volleys are completely (or almost completely) absent, because transmission is faulty in both directions. As for the remotely generated volleys, they are weak, and the local response to them is correspondingly weak

As can be seen, the cells of the nodes "can tell" when one of their links with the remote node is weakened, and can also tell whether the weakening is in the direction to or from the remote node.

Naturally, this statement does not say anything about the manner of the actual detection or storage of the information; it only points out that the necessary information can be deduced from the spike trains arriving to the nodes.

16.4 The Linkage Between Tracking, Metric Relations, and Long-Term Storage

The discussion of the *simple cell* in Chapter 12, describing a neatly arranged sequence of synapse pools which bring excitation from a gradually approaching sequence of contour waves (including the "near-miss synapses" and "synapses missing by a little more" in Figs. 12.1(a) and 12.2(a)), ignored the question of how the sequence might be created. A related question arises in connection with some of the upcoming sections (Sects. 16.5–16.6). The issue involves subcellular mechanisms which are beyond the scope of this book, but a few qualitative remarks are still in order.

It is most likely that the synapse pools which reflect gradual image shift arise during early learning, where contours repeatedly approaching the cell's receptive field are recorded and saved for future use. Such a possibility is all the more likely when one considers the literature on *reconnection* and *remapping* after damage to visual structures (Sperry, 1944; Chino et al., 1995), which suggests that visual cortical tissue, deprived of its input, can probably organize itself to process visual information from a place in the visual field new to it.

Here it can be pointed out that detecting successive snapshots of evolving reality amounts to *tracking*, and accordingly the most straightforward way to obtain a sequence of synapse sets which gradually approach a given version of the input (as if a *metric* existed on the synapse sets) is to record, during tracking, the incoming sequence of dynamically marked synapse sets (Sect. 13.2). Finally, if the sequence of synapse sets obtained through such tracking is to remain available many years later, when an adult uses tracking skills learned in infancy, the tracking must be able to access *long-term memory*, and be able to place much of its vast output of successive snapshots into it!

Because typical neurons are likely to participate in many tasks, it is legitimate to argue that memory formation must preserve selected synapse sets for later use, even after the same synapses are overwritten many times by unrelated long-term depression and long-term potentiation (LTD and LTP). Proper conditions must be able to bring them back into action again, and restore their grouping through the action of input recalling the one which originally formed the memory. When a synapse set is restored, it is expected that the restoration makes up for the fact that not all original members of the synapse set are excited in the later input, only "enough" of them are, including some linked through cross-potentiation; and that *completion* based on stored memories makes up for the incomplete input, possibly extending, for instance, a synapse set recognized on *one dendrite* to other dendrites to which it originally extended.

It may be stated that tracking probably improves with practice, but, as seen, an explanation of such improvement invokes a linkage between tracking and long-term storage.

16.5 Tracking a Contour Whose Shape Changes

Returning to the tracking of pairs of nodes, it will be noted that the visual apparatus is able to track images through changes which are much more disruptive than mere parallel shifting. The image of a flying bird, for instance, or of a waving tree, is *distorted* besides being shifted, which means that tracking must be able to cope both with distortion and parallel shifting of the image without the benefit of knowing which one is happening at any given moment.

In this context, "distortion" includes various transformations in addition to genuine distortion of the object itself, such as *rotation,* and it also includes 3D rotation, which changes the *perspective distortion* of the image.

When the image is merely parallel-shifted, the directional angles of its links do not change, and in the maintenance of tracking, the old cells can be replaced by new ones with the identical-same directional angles. When the image is distorted as well, the directional angles must also change. In order to maintain the links, the network must loosen its tolerances, which is the general first response to restore a weakened link, as described next, and then tighten them again by similar mechanisms as are used in the original linkup (Chapter 20).

16.6 Restoring a Weakened Link

As seen in Sect. 16.3, two nodes actively linked can detect when their linkage becomes weak. The weakening of a link means, by and large, that as the two nodes move, the axons from one node (or both) begin to miss the other node.

The task of deciding whether the target nodes are being missed because of distortion, or just inadequate adjustment to parallel shift, is likely to be beyond the capabilities of the primary visual cortex which implies that whatever measures the primary visual cortex may take to restore a link are expected to be designed to cover both possibilities.

As was seen in Fig. 16.1, the neurons at node 1 are in a position to distinguish between weakening of the 1→2 link (outgoing volleys) and the 2→1 link (incoming volleys) and take corrective measures for each case. It is assumed that a similar evaluation is made at node 2 with similar corrective steps.

The corrective measures are somewhat simpler when the weakened link is the one carrying *incoming* volleys, because then it is (sometimes) not necessary to add new neurons to the local ignitable group, only new synapses to its existing neurons, so as to bring input from a widened region at the remote location.

When, however, the *outgoing* volleys are detected as being weakened, it is never enough to add new synapses to the already active neurons, and new cells need also be added to the igniting pool, as shown in the illustrations of Fig. 16.2. The corrective measures are more or less the same whether the link is weakened as a result of

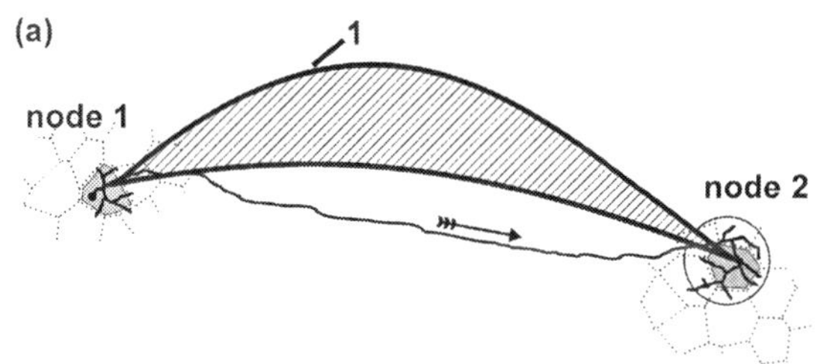

Fig. 16.2(a) A node widening its volleys to restore a link. Original form of the link. Two nodes (1 and 2) are shown together at two corners of a shape (shaded, marked "1"), and one of the axons in the link 1→2 arborizing around both nodes. The branches in node 1 support the ignitions which send volleys toward node 2.

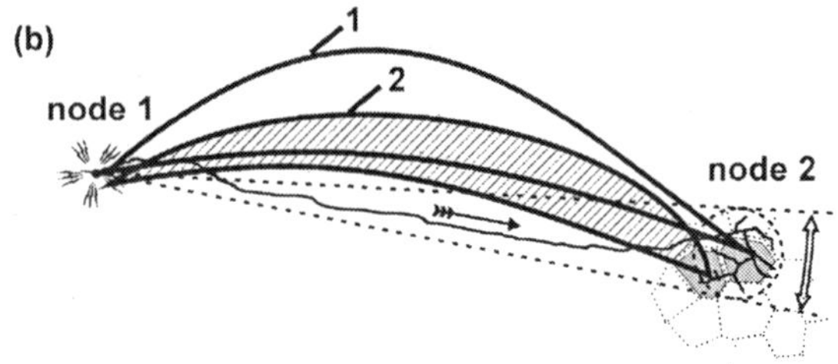

Fig. 16.2(b) A node widening its volleys to restore a link (Cont.). Shape is displaced and distorted, diminishing the 1→2 signal. The shape is changed into another (marked 2, shaded). The corner at node 1 still falls within the same column but corner at node 2 falls within a new nearby column. Further, the distortion is such as to change the directional angle required for 1→2 reaching

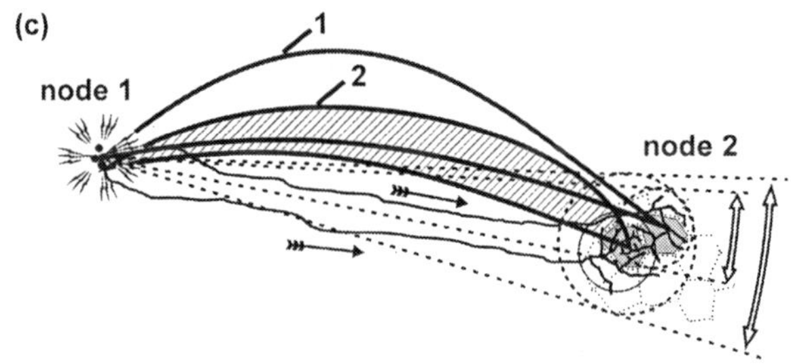

Fig. 16.2(c) A node widening its volleys to restore a link (Cont.). Node 1 recruits new cells. New cells are added to the bridgehead at node 1, by a reciprocity criterion, which results in tentatively sending volleys into a region wider than the original node, and displaced to be centered at the new source of volleys (*larger circle*).

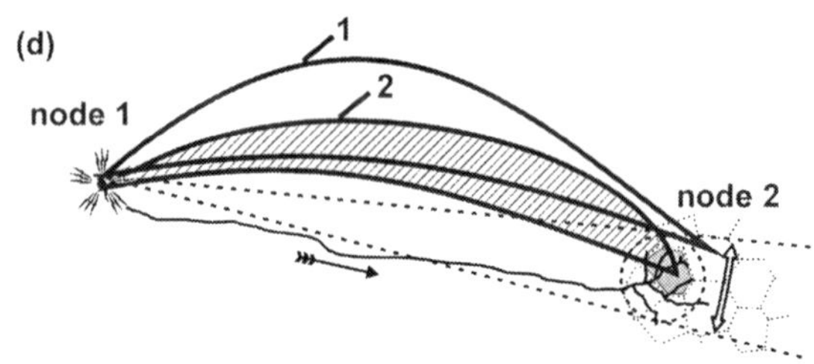

Fig. 16.2(d) A node widening its volleys to restore a link (Cont.). Cells failing to reach the new node 2 are later dropped. With the help of multi-column mechanisms (Fig. 17.1 (h)) and echo detection (Chapter 20, for instance Fig. 20.2i), the excess axons are eventually dropped and the range of directional angles returns to its former sharpness

distortion or parallel shift beyond expectation; in each case the bridgehead must widen the angle (and depth) into which its axons are sent. In other words (in this model), the corrective measure is to add cells to the current bridgehead, so it covers a wider range of angles and distances than before. The purpose is that the new and larger "effective field" will catch up with the elusive remote node, and have a sufficiently large overlap with it.

When new cells need to be added, cells throughout a widened range need to be prepared for playing each of the necessary roles in the node. They become "understudies," in the sense that they absorb all information necessary for playing a role in the node but will not necessarily be called upon to play it. As was seen in connection with the simple cells (Figs 12.1, 12.2, and 14.1), the preparation cannot anticipate the exact course of image drift, and cells only find out whether their preparation was needed when (and if) the contour moves into their receptive field. More will be said about this in connection with Fig. 18.1.

It will be noted that in all cases the corrective action makes use of an assumption of built-in local knowledge of the remote neighborhood relations, in the sense that at the input end the neurons must know which of their synapses bring input from "slightly" beyond the neurons feeding the current synapses. In effect, each neuron must have a "metric" on its set of synapse sets in a way analogous to the organization of the synapses on simple cells (Sec 12.5). None of this is trivial, since the location of synapses on the dendrites has no relation to the distant locations of the cells originating the synapses.

Similarly, to correct a weakening of outgoing volleys, the node must "know" which local neurons have *effective fields* just slightly beyond those of the currently active bridgehead. In other words, this kind of corrective action is an example of an operation which may benefit from early learning as suggested in Sect. 16.5.

The cells which are in the vicinity of the node 1 bridgehead but are not initially part of the bridgehead are the candidates to be added to the bridgehead in the attempt to restore the linkage. These cells are in a position to detect that the link to node 2 has become weaker (through the test shown in Fig. 16.1(d)), and know that they are to stand by to join the bridgehead. The decision whether to join is based on detecting that they are reached from the node 2 volleys in sync with the node 1 volleys. The situation is as in Fig. 18.1 below, where the issue is identical to the one here; it is also a decision whether to join a bridgehead actively linked to another remote one. Because of reciprocity, the cells reached by the igniting remote cells are likely to reach them in turn, and they are tentatively added to the bridgehead. The reaching is less accurate than the one from the existing node; the new cells send volleys into a wider region.

Chapter 17
Why Is the Drifting Retinal Image Helpful in Perception?

It might appear to the casual observer that the movement of retinal images should hinder vision, because it denies the system the time it needs to recognize images; images must be recognized before they move to some new set of cells.

But in fact it appears that the visual system manages to use the image movement to its advantage. Artificial stabilization of the retinal image interferes with vision (Riggs and Ratliff, 1952; Martinez-Conde et al., 2004).

The secret is that nodes can *grow* in the course of image drift, and by growing, acquire more long axons and thus have greater probability of reaching other remote nodes.

Further, the nodes cannot grow if the image does not drift. The reason is that an actual *contour* is needed for properly training the cells; the promise of a contour cannot do it. The node can blindly add neighboring cells and improve its chances of reaching, but there are also some further steps which require firing by the cells, and other events of interactive training, which can only occur when the cells participate in actions enabled by the presence of a contour.

Then, once a cell is recruited to a node with the help of a contour, tested for proper functioning, and trained to participate in all necessary ignitions, it is not a great error to retain it for a while after the contour has moved away. Its slight misalignment can be overlooked on account of the fact that it helps the node reach its target node.

It may be noted that the need for larger nodes usually arises when the target nodes which they must reach are located farther away, and this tends to keep the angular error roughly constant.

In the linkup procedure (Figs. 19.4 and 20.2) there is a phase where pairs of cell groups, one group in each of the nodes marked for linkup, repeatedly self-ignite while continuing to grow until they detect that their ignition elicits a response from the other end. Once reaching is achieved, further growth can be made to stop, as seen in Figs. 17.1(a)–(g). This makes the growth of the bridgeheads adaptive with the result that a node may contain some smaller bridgeheads and some larger ones.

C.R. Legéndy, *Circuits in the Brain*, DOI 10.1007/978-0-387-88849-1_17,

17.1 The Growth of Nodes in the Course of Contour Drift

The next series of drawings is meant as a rough illustration of the dynamics of node growth with the last drawing of the series, Fig. 17.1(h), showing a possible mechanism whereby growing bridgeheads can drop the columns no longer needed, so they retain more or less the same size even though they keep picking up new columns.

The groups can only grow gradually as the contour pulls through their columns, because each neuron added to a node must know that it is being added. The underlying contour waves, audible at any given time only at the momentary location of the contour, are part of the signal telling the neurons about the node (Fig. 18.1 below).

The idea behind the mechanism outlined in drawing (h) is that when a bridgehead is made up of a number of columns, its ignition is not instantaneous but a gradual spread of firing. This is because within a node, from column to column, the firing can only spread through the short unmyelinated (and therefore slow) axons. Accordingly, it is possible for the bridgehead to grow so large that by the time the local ignition reaches its far end, it has also reached the remote node (over fast fibers) and response has come back from the remote node (also over fast fibers). So, the cells at the far end of the local node receive the remote response to the local ignition before receiving news that there was a local ignition!

(a)

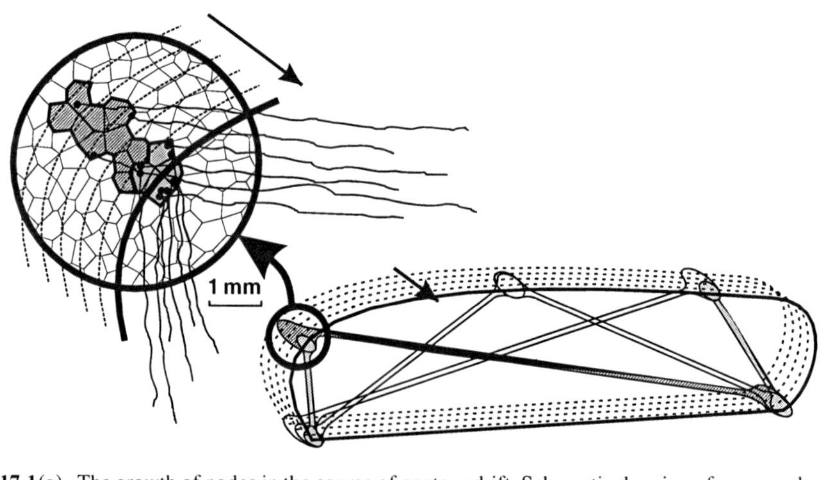

Fig. 17.1(a) The growth of nodes in the course of contour drift. Schematic drawing of some nodes and links. A contour with two corner nodes and three pass-through nodes is shown drifting in the direction of the *arrow*; *solid line* indicates its latest position. Long links are shown to have larger bridgeheads than short ones, in order to ensure reaching. The inset shows a magnified version of the circled node with the borders between cytochrome oxidase blob regions schematically sketched in. The beginning segments of some of the axons of two fiber tracts are added, with the local arborages left out. All bridgeheads of a node are shown as overlapping, each having grown from the same center (Fig. 18.2)

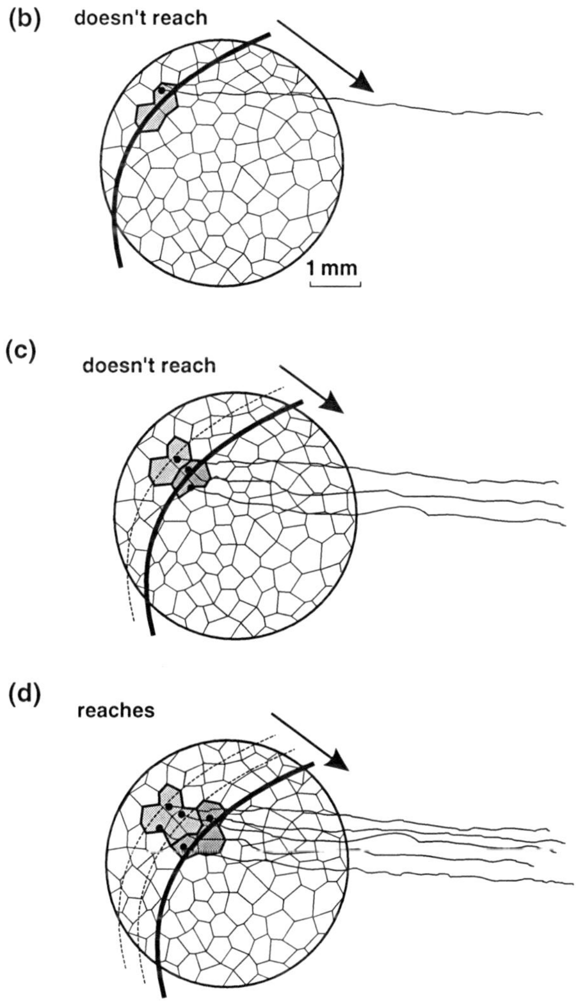

Fig. 17.1(b)–(d) The growth of nodes in the course of contour drift (Cont.). Cell group acquires columns as the contour drifts. Progress of a cell group as it grows until it reaches the remote node and becomes a bridgehead of a link. At the momentary location of the contour (*heavier line*), where the contour waves directly impinge on the cells, self-ignitions are initiated (Fig. 18.2). Ignition support links (not shown) between the bridgehead cells, formed while a column is "hot," are retained after the contour moves on (*lighter shaded area*). The events of this sequence occur during a phase of the linkup between nodes (Figs. 20.1(c) and 20.2(f)) where nodes emit volleys while also recording any incoming in-sync volleys

In the spike trains shown in drawing (h) as inputs arriving to neuron B (traces at right, top group), only the contribution from node 1 is shown (axons leaving at the right of inset at top); the spikes coming from local neurons are ignored. In the spike trains arriving to neuron A (middle and bottom group), the contributions from local

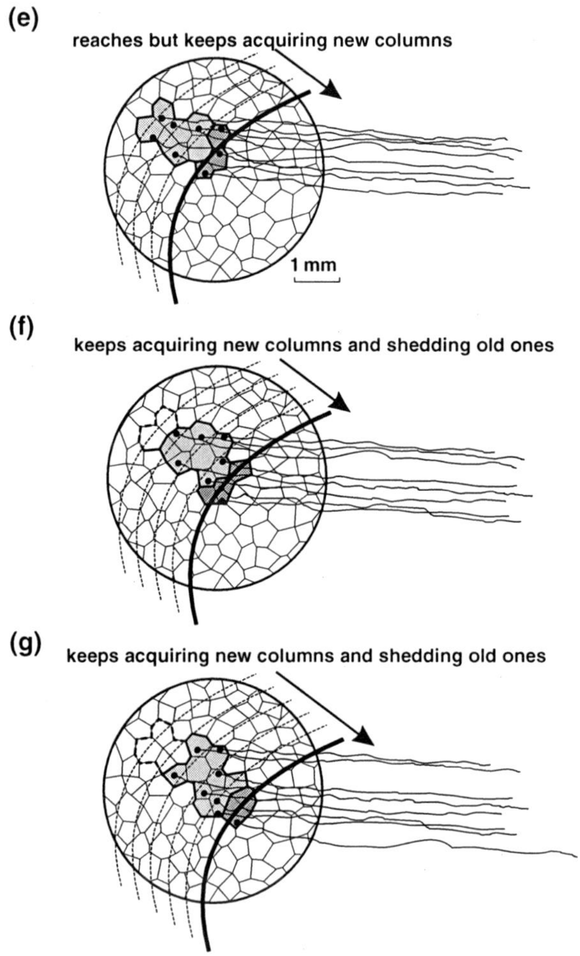

Fig. 17.1(e)–(g) The growth of nodes in the course of contour drift (Cont.). Cell group acquires new columns and sheds old ones. At some point, the node has grown large enough to reach the remote node even though it is quite far away. However, the growth of the node does not stop immediately; the node only stops growing when the timing of the echo from the remote node tells it to stop (see (h) below). After that, it begins to shed its oldest columns as it acquires new ones (see also Fig. 18.2(n))

neurons (node 1) and from the remote neurons (node 2) are both shown. The output of neuron B is shown as a separate trace with the top group; the output of A is not shown.

As seen in the spike trains originating in node 1, the ignition of the multi-column bridgehead spreads from column to column, usually via slow unmyelinated fibers. It is assumed here that the self-ignition of nodes starts from a "node center" (Fig. 18.2), located where the contour waves are running at the moment, and spreads

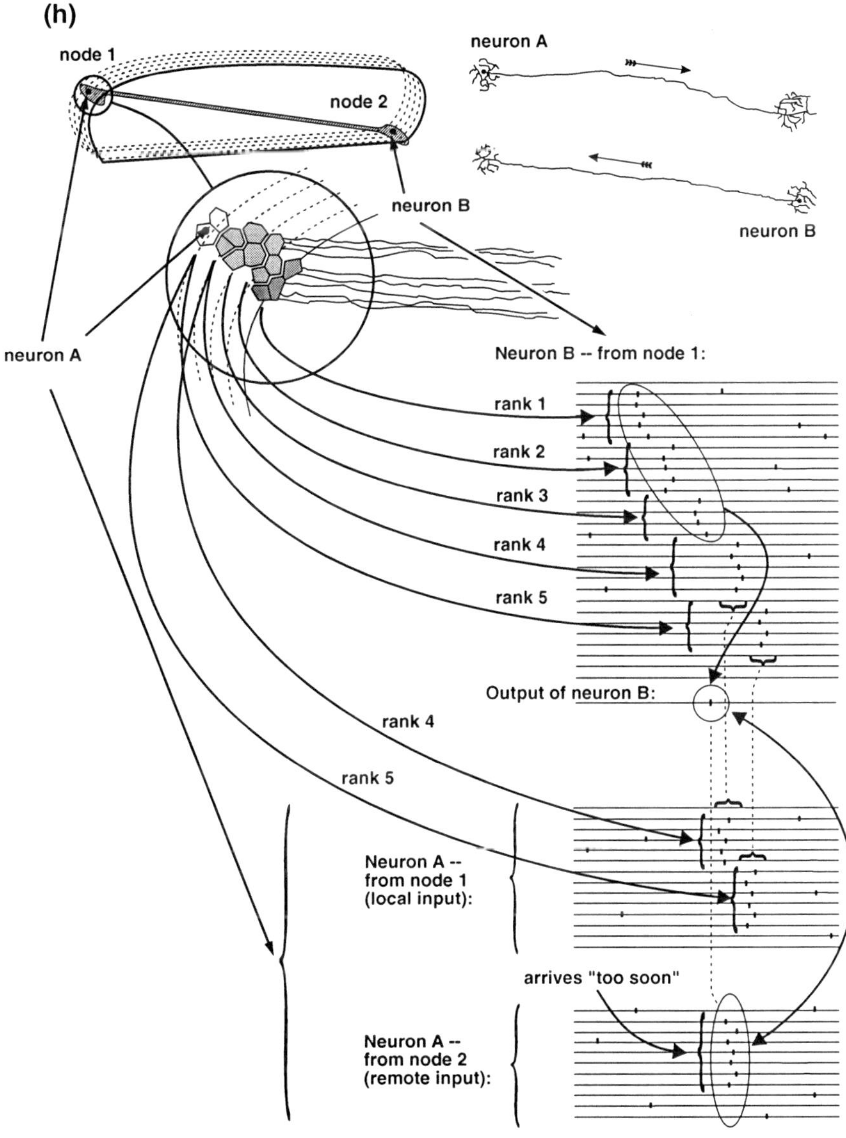

Fig. 17.1(h) The growth of nodes in the course of contour drift (Cont.). How neurons know when to quit a node. The drawing shows a pair of linked bridgeheads, one in node 1 and one in node 2, and the series of events initiated from node 1. The *top* traces show the spike trains arriving to a cell B in node 2, and the *bottom* traces the events, on the same time-scale, as received by cell A in node 1. The neuron A, situated in the oldest rank of node 1, is about to quit the node. The two nodes have a fairly long link connecting them (see (a)). Sketches at the upper right show stylized drawings of neurons 1 and 2 and their axonal arborages with Cajal-style arrows indicating the drome. The situation is the one envisioned in drawings (f) and (g), where the bridgehead in node 1 is large enough to reach the remote node and sheds columns as it gains new ones

from there to the rest of the columns by "ranks," sequentially from one set of columns to the next. A better (if less clear-cut) approximation to the situation, which describes two-dimensional spreading of the ignition, is shown in Fig. 18.2.

The test neuron A in node 1 is in the farthermost (and least recently added) rank of columns. During self-ignitions neuron A only receives input from its own rank of columns and the rank before it; it (by and large) does not receive input from the columns beyond those. In contrast, the connections from node to node are spread out more widely, and are without similar restriction; accordingly neuron A receives echo input from all parts of node 2 simultaneously and is reached by the echo volley as soon as enough of the ranks from node 1 have reached node 2 and received the echo from node 2.

In the illustration we assume that propagation over the long (Gennari) axons takes a relatively short time; in other words most delays in the spikes received by neuron B are due to the rank-to-rank passage of firing inside node 1. The drawing makes the whole volley look like a series of volleys, as the ranks of columns catch the firing one after the other.

What enables neuron A to know that it can quit the node is that neuron B (and the rest of the bridgehead in node 2) does not wait for all the partial volleys to arrive but fires as soon as it receives *enough* input to make its response.

The fact that node 2 responds before node 1 is finished with its drawn-out ignition means that neuron A receives the input from the local ignition "too late"; it receives response to the ignition from node 2 before it receives news of the fact that it is supposed to participate in a self-ignition.

This misplaced input, whose existence neuron A can detect quite clearly, is the signal to neuron A that it should quit the node.

Subsequently, A quits the node by disabling its ignition support links and its links from node 2 (synapse sets "1" and "2" in Fig. 16.1(a)). Similarly the other neurons in rank 5 drop out of the node (columns shown in white).

It is seen that the neurons of a column are not dropped as soon as there is two-way linkage, but a little later when they receive the late echo in evidence of two-way linkage. The system has a safety feature built into it against loss of linkage, making the links more "sturdy." Further, if the linkage is lost, columns stop being dropped, and accordingly the node keeps growing.

17.2 Kernel Cells in Multi-column Nodes

The thing that holds all the bridgeheads together, broadcasting that they are all in the same node, is the *kernel* of the node (Sect. 9.5), made up of *kernel cells* distributed throughout the node. The kernel cells send excitatory contacts to other kernel cells of the node, and to every bridgehead within the node, but not to cells outside the node.

The kernel cells, like the bridgehead cells, drift with the node as the retinal image drifts, and the kernel grows as the node does. In order to keep the upstairs informed of the node membership of the bridgeheads, it is necessary for them to refresh the identity information frequently. By the same reasoning which tells us that pairs of linked bridgeheads must repeatedly co-ignite to be tracked, a kernel must also repeatedly co-ignite with every bridgehead on its node.

A side effect of the interaction between co-ignitions is that co-ignitions of the kernel, the *vertical ignitions*, will often be incomplete, because some of the bridgeheads will not have recovered from their last previous horizontal ignition when called upon to co-ignite with the kernel; but sooner or later all bridgeheads are expected to be joined to the node. Some horizontal co-ignitions will fail because of failure of the secondary node to follow on the heels of a recent vertical co-ignition (as illustrated in Fig. 9.11), but once again the failure of a large number of ignitions will not disrupt the data flow as long as enough ignitions are successful.

Because of the mutual passive following between facing bridgeheads, a bridgehead in a node A, when following its local kernel in a vertical ignition, will often pass on its ignitions to the facing conjugate bridgehead in node B, since the latter cannot tell the difference between a self-ignition of the node A bridgehead and a volley emitted by he same bridgehead when it passively follows its local kernel. But that does not corrupt the data transfer either, since there are enough other times when the horizontal co-ignition is clean.

Since the kernel cells are short-axon cells, their excitation goes at most as far as the neighboring column. In nodes having many columns, the kernel ignitions spread gradually from column to column, and in each column they must pass on their ignition to the local cells of all bridgeheads.

Chapter 18
The Maintenance of Moving Nodes and Bridgeheads

In the sections dealing with simple cells and complex cells, we introduced the concept of tracking, along with the need to maintain "noun" objects, the kind made up of ever-changing sets of neurons, in a form suitable for tracking.

When the task is to maintain the cells that make up a bridgehead, there is the extra twist that all new cells added must fit in with two predetermined networks located in cortical places far apart.

18.1 Adding New Neurons to a Drifting Node

As was seen in Fig. 17.1, bridgeheads have to keep adding new columns as the contour drifts, and shedding old ones. Adding a column means, before anything else, telling the neurons in the column that they may be added to a bridgehead of a node.

As in the warm-up of simple cells (Chapter 12), this is made possible by the fact that when a neuron receives patterned input of some sort, the neurons nearby, by and large, also receive it. In this case, the result is that a cell is in a good position to join a bridgehead if it detects being reached both by ignitions from nearby cells and by volleys from distant cells time-linked with them. The remote input is an indication (although not assurance) that the cell is of the right kind to reach the remote cells sending the volleys, because of the general principle of (more or less) reciprocal linkage. The only thing holding the cell back from firing along with the local ignitions is that it is not on the contour, which it knows from the fact that it does not receive local contour wave input. Once it does, it joins.

Schematically speaking, it can be said that the process of adding a neuron is made up of two phases; in the first the candidate neuron does not fire, and in the second it does. Thanks to the repeated co-ignitions in an *active link*, it is possible (in the first phase) to select the neurons reached by both nodes without need for those neurons to participate in the ignitions (Fig. 18.1(a)).

Once those neurons are selected, it is necessary (in the second phase) to interconnect the suitable neurons into an ignitable group (Fig. 18.1(b)) and then to verify that the ignitions of the group can reach the other node (Fig. 18.1(c)).

C.R. Legéndy, *Circuits in the Brain*, DOI 10.1007/978-0-387-88849-1_18,

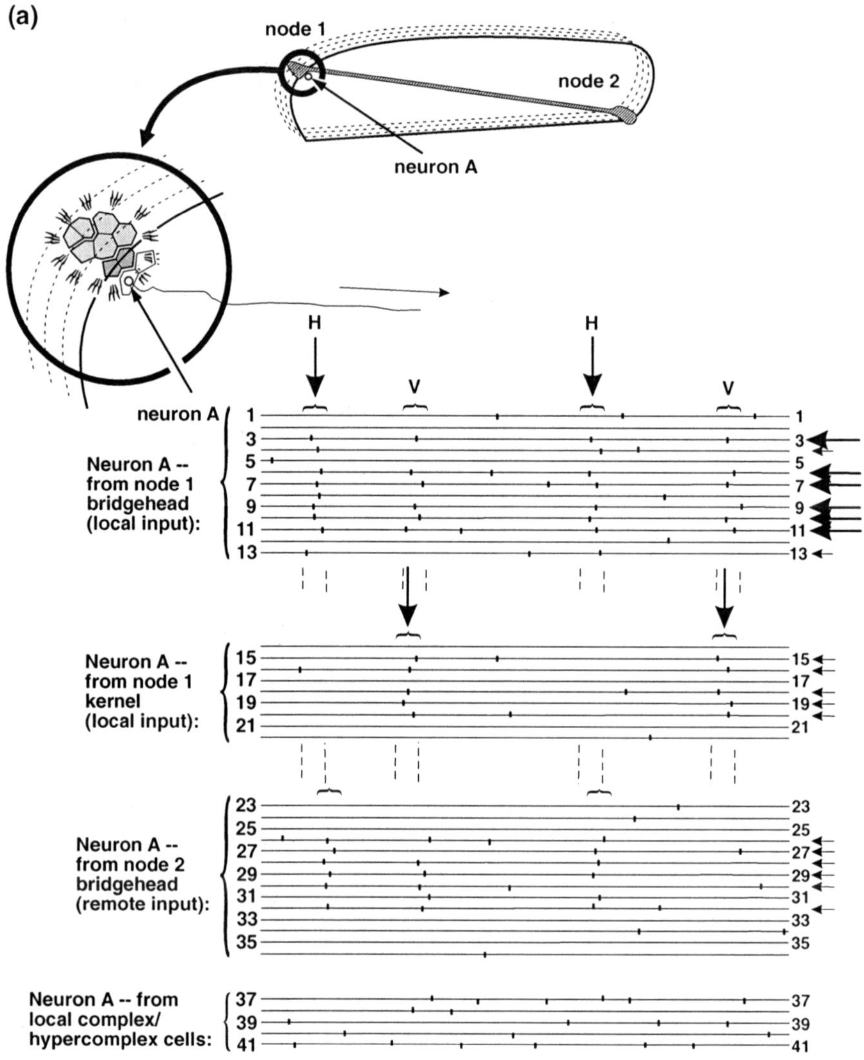

Fig. 18.1(a) Adding new neurons to a drifting bridgehead. Marking the synapses on nearby cells reached by both nodes. Separate sets of traces show the input to a neuron from four input sources: from the local bridgehead (node 1), the local kernel (node 1), the remote bridgehead (node 2), and the local complex/hypercomplex cells. The volleys initiated from the remote node (node 2), and the local responses to them, are not shown. *Arrows* on top show the neuron set initiating the co-ignitions; two co-ignitions are initiated by the local bridgehead cells (horizontal co-ignitions, "H") and two from the local kernel cells (vertical co-ignitions, "V"). During the interval of time shown in this first drawing the contour is approaching neuron A but has not yet reached it, as seen in the bottom traces, which do not show evidence of contour waves

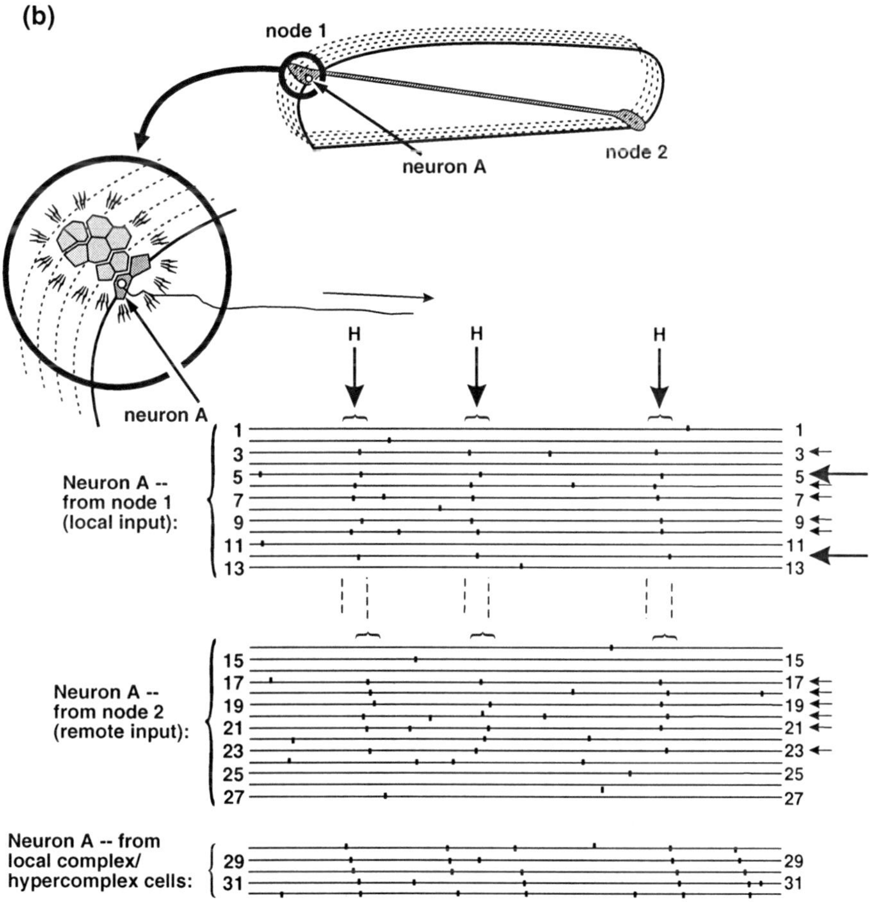

Fig. 18.1(b) Adding new neurons to a drifting bridgehead (Cont.). Linking the marked cells into an ignitable group. In this drawing the contour waves reach the neuron A (bottom traces contain runs of contour waves). Accordingly, the neuron can start participating in all the co-ignitions, horizontal and vertical (only the horizontal ignitions are shown. The role of kernel cells and of vertical ignitions was completed in step (a))

Neuron A in Fig. 18.1 is chosen to be one reached by the occasional co-ignitions between the local (node 1) bridgehead and node 2 (horizontal ignitions), and also reached by the occasional co-ignitions between the same local bridgehead and the local kernel cells (vertical ignitions).

After a few repetitions the neuron marks the synapses on which the volleys arrive. It will be noted that in the top traces some cells participating in the horizontal ignitions do not participate in the vertical ignitions (small arrows at right). Only the ones participating in both ignitions have their synapses marked on new candidate neuron A (larger arrows), and A is eliminated as a candidate as the number of these is too small.

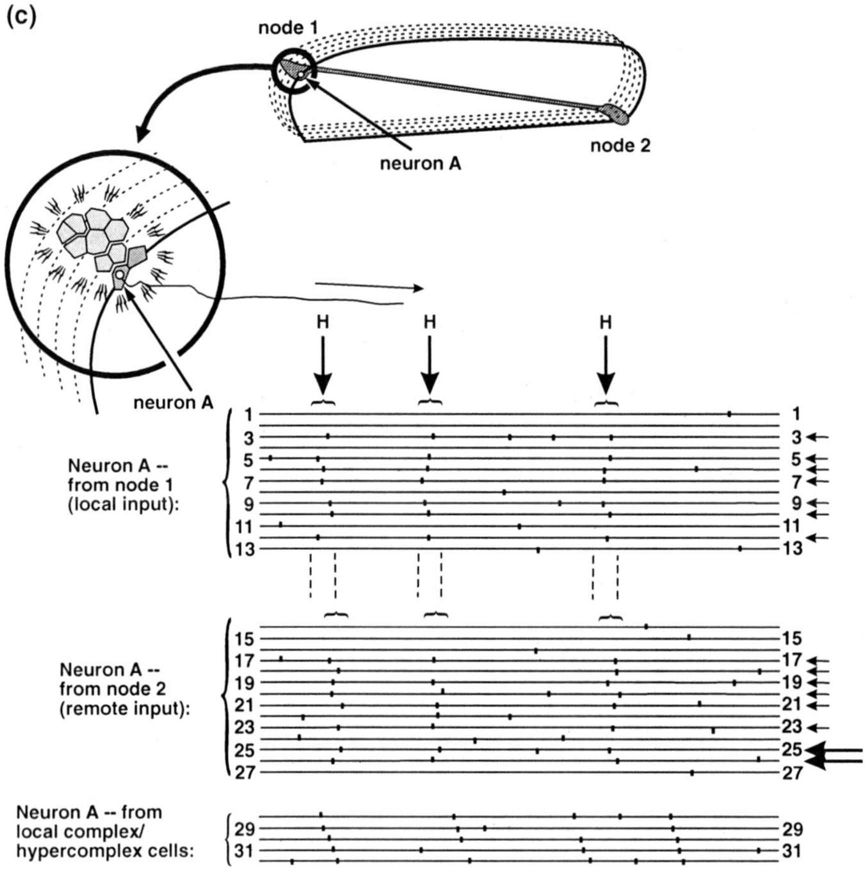

Fig. 18.1(c) Adding new neurons to a drifting bridgehead (Cont.). Selecting neurons able to reach the remote node. Next, the difference between the original response and recently changed response from the remote node is evaluated. If there is no change, it means that neuron A's participation did nothing to strengthen the remote response, and neuron A is removed from the node, its newly changed synapses returned to their former state

The requirement to drop bridgehead cells which are not reached by the kernel cells has the effect of preventing the different bridgeheads from drifting apart from one another as they evolve.

Since the set of candidate cells now fires, it is possible now to tie the cells of the newly added columns, on the basis of their firing, to each other. (In being allowed to fire with the rest of the cells, they have already been tied to previously active members of the ignitable group.)

Accordingly, it is now necessary to link up the newly active cells to the other newly active cells. This is done by detecting the difference between the volleys in (a) and the volleys in (b), and marking the synapses corresponding to the newly added cells; in the case of neuron A this means marking (and enhancing) the synapses from the cells corresponding to tracks indicated with the larger arrows on the right ("local input" traces 5 and 12). Through reinforcement of the synapses so marked, the new cells become active contributors to the subsequent local ignitions.

The first tests for node membership of neuron A have now been passed: the cell is reached by the local long-axon-cell ignitions, the local kernel ignitions, and the remote long-axon-cell ignitions; in addition it has, in step (b), detected the presence of the contour in its receptive field.

The objective of this step is to perform the remaining test: to decide whether the cell's participation helps node 1 in reaching node 2. This in turn is decided by testing whether the local volleys (in which neuron A now participates) enhance the response from node 2.

The problem facing the test is that neuron A is only a single cell; its output is weak, and a responding volley from the remote node may not be in response to the firing of neuron A itself but to the firing of other cells that fire in concert with it.

And in fact it is not possible to establish with certainty that neuron A itself reaches the remote node; however, it is possible to establish whether the newly added cells *together with the old ones*, and functioning as a coordinated group, can elicit a response from node 2, and add neurons to the responding set.

In a phase not included in the drawings, the cells of node 2, like those of node 1, detect the added neurons shown in (b), strengthen their synapses from them, to include them among the contacts causing node 2 to passively follow the ignitions of node 1. In this way, some node 2 cells will be brought beyond threshold by the newly added node 1 cells, and fire. It is these newly responding node 2 cells which are detected in (b) and (c).

The neuron A, and the others like it, can detect whether its participation extends the response from the remote node by looking for the difference its ignition made in the volleys arriving from the remote node.

The drawing (c) is made so as to show that the test neuron A does indeed receive the input which will confirm its addition to the bridgehead. The desired evidence is seen in "remote input" traces marked out with larger arrows at the right (traces 25, 26). The neurons shown in these traces did not participate in the response during (a), and now participate. The synapses from these neurons are now marked, and thanks to their presence the candidate neuron A is confirmed.

In steps not shown, a set of candidate kernel cells is also selected and developed by using similar criteria properly modified for their role.

18.2 Spread of a Bridgehead Sideways, Along the Contour

Before leaving the subject of the growth of nodes, it is necessary to point out one more twist: that nodes in fact probably grow in two dimensions rather than one. It therefore becomes necessary for the system to regulate the growth of nodes in the direction tangential to the contour, and limit that growth. It also becomes necessary to deal with the question that self-ignitions must always start in the same place on a node, and the place must be the same for all ignitable groups in the node (Fig18.2(e–o)).

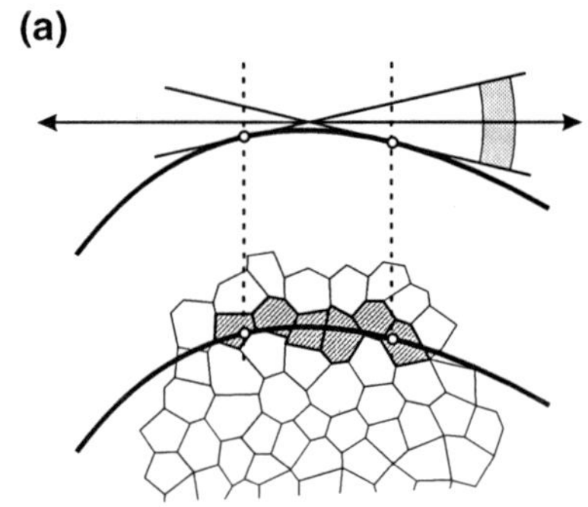

Fig. 18.2(a) Spread of a bridgehead sideways, along the contour. Initialization of a set of columns on the contour. In a slope-driven selection of node location the uncertainty in orientation selectivity will tend to select more columns than one; the smaller the curvature, the more the columns. The node size self-adjusts in the subsequent steps (see steps (m)–(o))

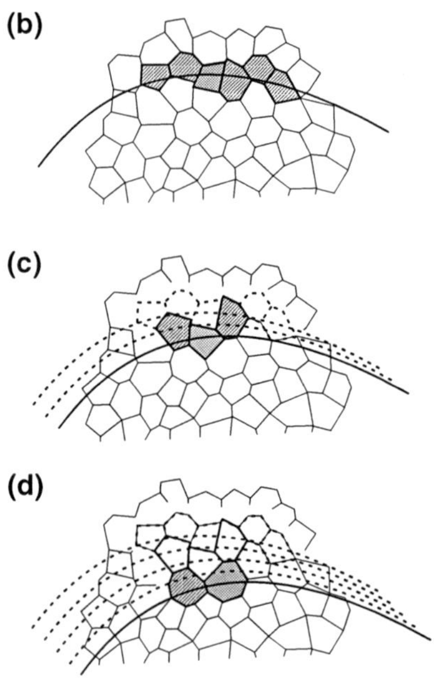

Fig. 18.2(b)–(d) Spread of a bridgehead sideways, along the contour (Cont.). Selection of a single column to serve as the node center. When the number of columns selected by the slope criterion exceeds one, it is necessary to select a single column which is to act as the prime mover for all self-ignitions in the node. Repeated instability then only exists in the one selected column, it initiates the self-ignitions and those spread from it to the rest of the node

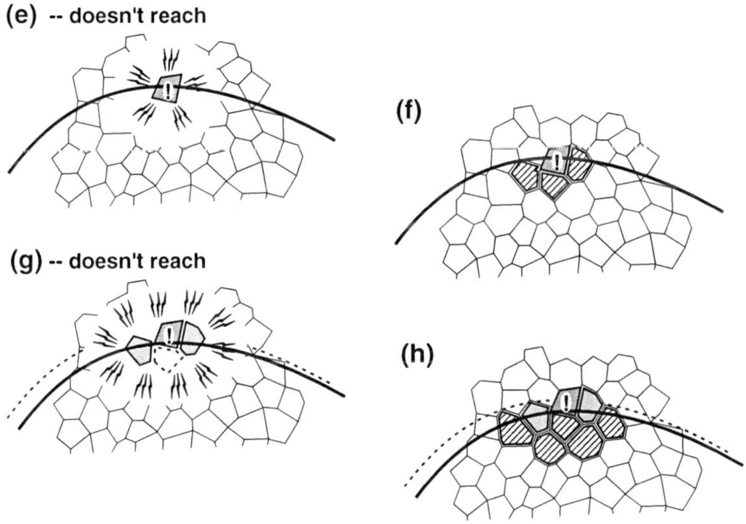

Fig. 18.2(e)–(h) Spread of a bridgehead sideways, along the contour (Cont.). Node center initiates ignitions of growing bridgehead. The node center initiates ignitions, and all neighboring columns on the contour tentatively respond. Since the remote node is not reached yet, and there is no possibility of an echo, the neighboring columns have no way of knowing whether their firing is useful. They therefore all continue firing along with the node center until they receive an echo and it tells them to stop

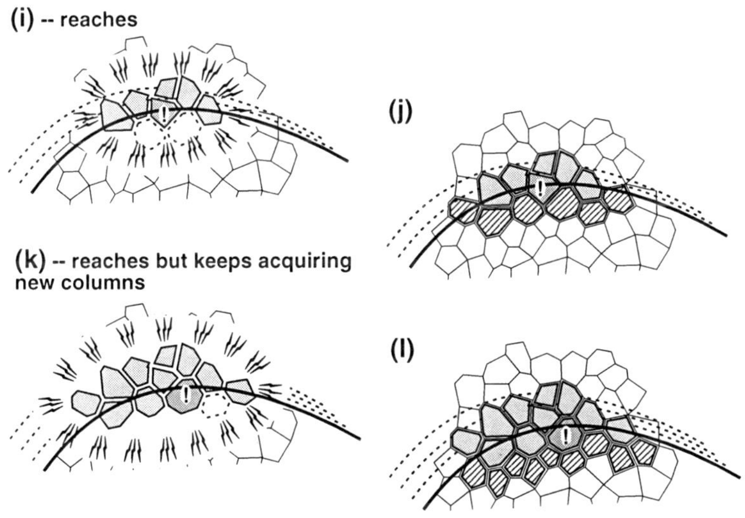

Fig. 18.2(i)–(l) Spread of a bridgehead sideways, along the contour (Cont.). Bridgehead reaches but keeps growing. Drawings (e)–(o) here are the two-dimensional versions of Fig. 17.1(b)–(g). The earlier "one-dimensional" drawings with their clear-cut "ranks" are better suited for illustration (Fig. 17.1(h)). The igniting set at each stage is shown in the drawings at left with the central column marked with "!" acting as the prime mover of all ignitions. The corresponding drawings at right show the "understudy" columns in the process of being warmed up (cross-hatched), to be activated when the contour enters them. In the subsequent left-hand drawings the columns previously prepared are shown participating in the igniting set, provided that the contour reaches them (prepared and not-yet-reached columns are left white)

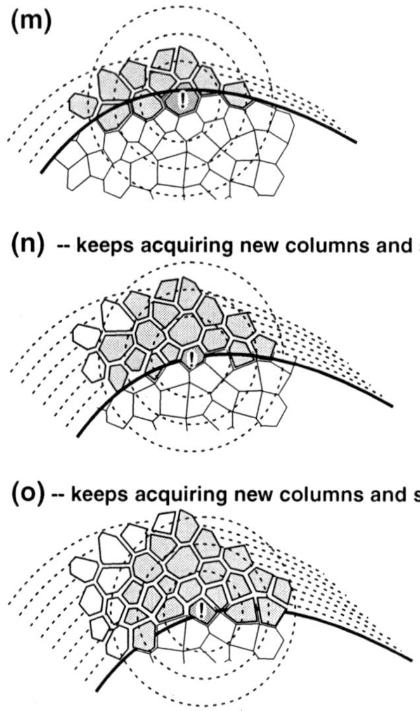

Fig. 18.2(m)–(o) Spread of a bridgehead sideways, along the contour (Cont.). Bridgehead sheds columns too far from ignition center. The mechanism for shedding columns, as described in Fig. 17.1 h, also works in a two-dimensional array of columns, as shown in these drawings, except that the "ranks" are roughly aligned on concentric *circles* drawn around the node center. As in Fig. 17.1 h, some of the newly active columns receive their echo volleys "too late," and are shed from the node (stop responding to the node center and undo their node-related synaptic changes)

The node center is the same for all bridgeheads and the kernel cells, and all self-ignitions start from it. Accordingly, all bridgeheads share at least one column: namely the node center. If the node were not organized around a single center, and self-ignitions could be initiated in two or more places, race conditions would arise and the protocol for dropping unnecessary columns (Fig. 17.1(h)) would fail.

Accordingly, at the beginning of node formation all the columns permitted by the slope criterion participate in a session of competitive selection aimed at deciding on a column to serve as node center (the details of the selection process are not addressed here), and once a node center is selected, regardless whether the selection is optimal, subsequent generations are selected from among its immediate neighbors, as shown in (c) and (d), each time repeating the competitive selection step.

In a corner node, the column containing the corner is automatically the node center, and there is no need for competition.

In the rest of this series of drawings the schematic sequence shown in Fig. 17.1 is repeated in two-dimensional illustrations arranged around the node center which starts all ignitions.

A system driven by a single node center, shared by all bridgeheads and the kernel, automatically ensures that all these groups (usually having different sizes) share at least one column. The shared column supports the mutual reaching needed in vertical ignitions.

In addition, the nodes spreading in two dimensions from a single node center remain spatially concentrated, more so than the long and narrow nodes of Fig. 17.1, and in general lead to more accurate angle measurements.

It will be noted how the "node center" system makes sure that the different bridgeheads, developing in roughly concentric semicircles, overlap optimally, to reach one another via short axons.

Part IV
Firing Games and the Integration of Contours

Chapter 19
Making the First Links by Crawling Along a Contour String

In the end of Part I of this book (Chapter 9), it was shown how co-ignitions could describe shapes, but the question was left open as to the way the co-igniting groups might be assembled on the basis of the visual image. Then in Part II the contour string was described as a temporary avenue of activity arising whenever a contour appears on the retina. Its signals, in the form of complex and hypercomplex cell outputs, can tell other nearby cells when a contour is present in their receptive field. In Part III the nodes were added, to stand in for the points linked by co-ignitions, and it was shown how ever-newer cells could be recruited to them by methods utilizing the drifting of retinal images. In addition it was seen that nodes, once linked up with other nodes, could execute repeated co-ignitions, continually broadcasting their status and momentary relations.

The piece left to last was the way in which nodes might find each other and form links. As will be seen, this piece is tricky because of the limited information initially available to neurons.

The links of interest to this model are all between nodes on the same contour, because the general objective is to convey the shapes of contours through co-ignitions of the nodes on them. Only if the nodes are on the same contour will the totality of their co-ignitions characterize the shape of a contour.

In the sequence of events at the beginning of a fixation, the first links formed between nodes are the *contour links*, because the first links must by necessity be formed by crawling along the contour in some way, from node to node.

In the pages below, much of the discussion will revolve around deciding whether nodes are on the same contour as other nodes. The formation of the *contour links* (the subject of this chapter) will be seen to be the more difficult task in this regard than the formation of the longer links (the subject of the next chapter), because once a set of contour links have been created with a guarantee that they connect nodes of the same contour, it is easier to form additional links by combining them, and retaining the same guarantee.

C.R. Legéndy, *Circuits in the Brain*, DOI 10.1007/978-0-387-88849-1_19,

19.1 Outline of the Continuity Detection and Contour Linkup

In the next series of subsections (and the similar series in 20.1) I will outline verbally the heuristic arguments underlying the sequences subsequently shown in figures. The purpose is to spell out the issues related to the "logistics" of local knowledge, the system's way of making sure that the knowledge is available at the localities when it is needed there. (Some of the same points are raised in these sections as in the later figures, and the reader may well prefer to skip either the sections or the figures, here as well as in Chapter 20. I included both, because I found the dual perspective helpful.)

19.1.1 Nodes and Their Initial Ignitions

As was stated in Sect. 15.4, the nodes on contour strings in essence play the role of points on the contours, for the purposes of determining the slopes of lines connecting them. Since the nodes are meant to single out locations on a contour, the contour string must form before its nodes can form. Then, since the nodes need to make themselves known to each other before they can interact, they must first develop their ability to ignite individually, then make contact with each other in preparation for igniting together.

In the interval between the time when they start their self-ignitions and the time when they start linking up, the nodes, through their ignitions, announce their presence to all cells nearby, in particular to the cells of interest here, the ones on contour strings. The cells receiving the repeated volleys, some of them on nearby nodes, others on the segments between nodes, are in a position to mark the synapses bringing the volleys (Fig. 19.4(b, c)) as ones coming from nodes, since only nodes send out ignitions on Gennari fibers. But they do not know whether the nodes are on the same contour as they themselves, or some other nearby contour.

After marking the sets of synapses bringing ignitions, the next task for the nodes is to determine which synapse sets bring volleys from the same contour.

19.1.2 The Cells as Individuals Cannot See the Whole Picture

So, the question is, how can a node find out that it is on the same contour as another node? The solution given here makes use of the properties of contour strings as developed in Part II, and the properties of the nodes as developed in Part III. Accordingly, it is desirable to take a new look at the properties of contour strings and nodes, and examine them from the point of view of local knowledge and its flow.

The problem is that the retinal image of a contour stimulates many cells, independently of one another, and does not tell the cells about all the other cells it also stimulates. And in particular the fact that one node is linked to another node by an uninterrupted contour is not known to either of the nodes.

Between them, the many cells which lie between the nodes know all there is to know about the contour and its uninterrupted course, but the crucial thing to remember is that only these cells *together* are in a position to tell others about the continuity, and in order to act together these cells need to form a cooperative group. Only through cooperation can the cells send a coordinated message dealing with the over-all properties of the contour (see 10.4).

In order to create coordinated output, the cells of the group must form functional synaptic links with one another, and to do so they must know about at least some of the other cells which are to be part of the cooperative group. This is accomplished by formation of what we called the *contour string*, and the synaptic relations between its cells. Thanks to these relations, and in particular the *trigger synapses*, the simple cells making up the string are in a position to know that at least a few other simple cells near them have receptive fields on the same contour.

19.1.3 How a "Grand Design" Enables Cells to Convey More Than They Know

Now, the linkage enables the co-excited simple cells to send out the continuity information to the rest of the brain, but the cells involved have no assurance that their shared excitation has its roots in any fact of objective reality. They rely on the over-all design of the system for their assurance.

When the image of an object falls on the retina, it gives rise to an unusual constellation of visual details, in this case one where many similar edge elements are lined up end to end. With great likelihood, it is a constellation that indicates some real-world object, for it would have a low probability of coming together if the scene were a random salt-and-pepper pattern. The over-all design converts the unusual configuration into unusual events by temporarily tying together the simple cells responding to adjacent edge elements.

Thanks to the temporary linkage, as seen in Part II, the chain of simple cells together manages to create a firing pattern of macroscopic dimensions, which can be utilized by the rest of the network to make further recognitions. The larger pattern arises by way of the propagating wave created by the rules governing the cells in their chained mode, and the chained mode is entered thanks to the ability of the cells to recognize, based on locally available signals, the times when they are to enter it.

This illustrates the way in which the "grand design" of the system can enable cells with no knowledge extending beyond their locality to get together and convey information which imparts knowledge regarding the global properties of a pattern.

19.1.4 Localities Monitoring the Moving Wave Via Long Axons

The fact that a propagating wave can move continuously over a length of contour solves one part of the problem.

But there still remains another part, the problem of telling any one cell that it is on the same contour as some other cell a long distance (many millimeters) away from it. Here, a solution requires that knowledge originally *distributed* among many neurons be first gathered and *concentrated* in one neuron (or, to be more useful, in each member of a group of neurons).

A solution offers itself through the property of contour waves that they "hug" a line of co-excited cells, and die if the contour is interrupted. If a contour wave starting from one neuron can make it all the way to another neuron, it follows that the two neurons are on the same contour. The only question is how one neuron can be made to know whether a contour wave manages to move from it to another far-away neuron, given that it cannot "see" the contour wave as it moves.

It is at this point that the stria of Gennari enters the picture with its massive supply of long axons. While the cell cannot "see" the wave moving down the line, it can be continually kept informed of its passage, provided that long-axon neurons are available everywhere where the contour wave passes, with their long axons reaching back to the neuron which has to be informed of the passage of the wave. All that is then needed is a way to arrange that the long-axon cells in question emit spikes wherever the contour wave passes.

In other words, a system of long-axon cells, if successfully made to generate a wave of activity which shadows the contour wave, can permit a target cell reached by all the axons to monitor the progress of the contour wave, and *reduce the problem of detecting spatial continuity to a problem of detecting temporal continuity*. The process is illustrated in Fig. 19.3 below.

19.1.5 The Smallest Cell Group Able to Trade Knowledge: The Node

In the last few paragraphs we pretended, for the sake of illustration, that a single neuron was in a position to know that some other single neuron was on the same contour, but here a practical problem must be taken into account: the problem of *reaching*.

If fixed cell groups are to be carriers of knowledge needed by the rest of the system, the groups cannot be too small. They must be large enough to be reached by all groups from which they may need to acquire knowledge, and large enough also to reach all other groups to which they may potentially need to pass on knowledge.

These requirements are the same, only said in different words, as the requirements stated in introducing *nodes* in Sect. 15.4. Nodes can be thought of as being the smallest units where knowledge regarding spatial relations, a kind which is on interest in this model, can be concentrated. The requirement of reaching other nodes, including nodes far away, is the reason node sizes need to be adaptively determined (Fig. 17.1).

Because nodes are the smallest unit in which knowledge can be usefully concentrated, it is necessary to arrange that knowledge is also *shared* among neurons within nodes (see 16.3), meaning that the same knowledge is accessible to any "typical" member of the same node (see for instance Fig. 16.1). In practice most knowledge is only imparted to the bridgehead dealing with it, but when necessary the knowledge can be spread to the rest of node (see below); therefore it can be legitimately said that the *node* knows what its various neuron groups know.

This means that the task of making a neuron know that it is on the same contour as another neuron should be restated as the task of making *one node* know that it is on the same contour as *another node*.

In the linkup sequences in this chapter, the node which "monitors" the contour wave through long axons will be referred to as the *base node* of the continuity detection.

The drift of retinal images raises an added problem in the description of nodes: the sharing of knowledge between all cells in a node must keep up with the continual change in the membership of the node. As new neurons (and whole new columns) are added to a node, they need to be taught all that the old neurons know.

The way in which the new members of a node learn what the old members know was discussed in Figs. 18.1 and 18.2. They learn it from the same sources as did the old members; from the volleys received from groups of axons together with the chemical signature of the axons bringing the volleys.

19.1.6 Monitoring Single Contour Waves in Isolation: The "Tracer Wave"

It may at first seem reasonable that continuity of a contour could be ascertained at the base node if reports of the moving contour wave could be continually sent back to the base node, putting the base node in a position to detect gaps in the arriving shower of pulses.

The problem with this simple system is that contour waves are generated so frequently (see Fig. 10.3) that most of the time several of them are in progress between any two nodes. Even if all contour waves were diverted or stopped by some gap before they reached the next node, the remaining contour waves would mask the break.

The solution is to devise a protocol whereby the neurons between nodes can select a single contour wave, and follow it to the exclusion of all others that go before and after it, until a clean message is received at the base node. This means that the long-axon neurons must not blindly shadow all contour waves but only follow the ones they are told to shadow.

The next question is how will all the neurons in the chain between the nodes know which contour wave to follow?

The first iteration of the answer is that since the base node must be in a position to monitor the results, it must also be in a position to initialize the wave-following axons. The initializing message must by necessity be a volley emitted from the base node, telling all the long-axon neurons it reaches that when the next contour wave passes them they are to fire once, then ignore all other contour waves after that until the next volley. I will refer to this volley by the name "*tracer initialization*" (*tracer init*) volley.

If the Gennari cells respond to the next contour wave they receive after the *tracer init* volley, the result will be a single wave of pulses, shadowing the movement of the next contour wave, and detectable at the base node.

19.1.7 Preventing Extra Waves from Being Traced: The "Second Enable"

There still remains a problem which must be solved before the *tracer init* can isolate a single contour wave. If "the next contour wave" after the *tracer init* is not the same one for all the cells to the right of the base node, the *tracer init* protocol will not yield a clean tracing of contour waves.

In particular, if there happen to be any contour waves already in progress from the base node to the next node at the moment when the *tracer init* is emitted, those contour waves will also be traced and, will compete with the contour wave we want to trace.

The solution we use is to include one more element in preparation for tracing, additional to the *tracer init* volley, in the form of a "*second enable*" requirement which only lets the long-axon neurons fire as part of a tracer wave if the long-axon neurons right before them have also fired as part of a same tracer wave, except in the region immediately after the base node, where *second enable* input is the *tracer init* volley itself (Fig. 19.2(b)). Since this means that every "second enable wave" by necessity starts at the base node, when a contour wave is already in progress (away from the base node) at the time of the tracer init, it will get its initialization, but not its *second enable*.

It will be noted that the insistence of the doubly enabled tracer waves to start from the base node initiating them is also helpful in avoiding crosstalk between contours; in other words the problem that volleys from another nearby contour can sometimes be mistaken for a tracer init (Fig. 19.6).

19.1.8 Satisfying the Surprise Requirement of "Second Enable": Warmup Runs

This would ensure continuous tracing, but a problem must still be solved, concerning the surprise requirement of the *second enable* input. The problem is that if tracer waves cannot get through without a *second enable* input, the cells for which the input is intended will not receive the preliminary firing they need to train (mark) the synapses bringing the *second enable*.

The solution used here is to add, before the start of full-fledged tracer action, a few preliminary (warm-up) runs in which the *second enable* requirement for tracer report firing is dropped. During the warmup runs the synapses can be marked. The marking will be correct, since the warmup waves include the correct waves which are to survive to the later runs, and those waves travel through the same neurons and same synapses as in the later runs; the only difference is that warmup runs include other waves which later drop out.

Cells exit the warm-up mode as soon as they receive enough surprise from their newly marked second-enable synapses to start requiring their input.

The next question concerns the neurons which are supposed to carry the tracer waves. How are they to distinguish the *tracer init* volleys they are supposed to follow from the similar-looking volleys from other contours which they are supposed to ignore?

The answer is that in the warmup runs they cannot distinguish the two and will in fact run tracers for all volleys including ones from other contours. However, once the warmup is over and tracer reports are no longer emitted without a *second enable*, the volleys received from wrong contours will not be supported by a *second enable*, because tracer waves (and their accompanying *second enable*) cannot jump from one contour to another. In that way the cells are in a position to remove the synapses of wrong-contour volleys from their *tracer init* synapse sets.

In earlier drafts, I combined the creation of nodes with the tracer waves. Starting out from a corner node the earlier system created a new node right before the tracer reports became too weak to reach the base node, then repeated the process and repeatedly created pass-through nodes until the next corner node was reached. Later, I abandoned the solution for the quicker and more consistent one shown here.

19.1.9 Tracer Waves Continuous with an "Arrival Volley" from the Next Node

The tracer waves shadowing a contour wave, and sending back their tracer reports as they move along, will tell the base node (let's call it node A) about gaps in the contour, but (in the form described so far) do not tell it whether the uninterrupted contour extends all the way to any particular other node.

Therefore what is still needed is a volley (*arrival volley*) coming from the next node on the contour (let's call it node B), emitted when the tracer wave reaches that node from node A, and therefore continuously following upon the tracer shower, without any pause in-between (Fig. 19.3(b)).

The geometry of the situation is such that the *tracer wave* can be made to create an *arrival volley*, and thereby provide the needed assurance of continuity; all that is needed is that both the *tracer init* and the *second enable* act on the cells of node B as they do on the other long-axon cells reaching node A.

19.1.10 How Can the Base Node Recognize the Arrival Volleys?

One may ask, how can node A recognize the *arrival volley* as being the result of an *ignition* rather than just a somewhat louder segment of the tracer shower?

The answer is that node A repeatedly *receives* volleys from node B, which come to the same synapses as the arrival volleys. These arise because node B, like node A, is actively generating self-ignitions (to initialize tracers running oppositely to the tracers initiated by node A, Fig. 19.4). In this way the synapses on which the arrival volley reaches node A are already marked at node A, and are easy to distinguish from the rest of the tracer shower.

However, this also means that based on their synaptic pattern the arrival volleys are indistinguishable from the self-ignitions generated by node B. Therefore, in addition to synaptic pattern, the recognition of arrival volleys must make use of their timing. The self-ignitions of node B do not follow upon tracer waves starting from node A, but the arrival volleys do – provided they are not pre-empted by other mechanisms. To make sure they are not, the *tracer init* must suspend the autonomous self-ignitions of node B, until B emits its arrival volley.

19.1.11 Saving the Detected Continuity in the Form of Hardware

The *arrival volley*, in conjunction with the tracer waves using second enable input, provides a powerful method of continuity detection. While it was mentioned a moment ago that node A is in a position to monitor the tracer waves for gaps, such monitoring is in practice unnecessary, because if there are gaps, the tracer wave will never make it to node B and an *arrival volley* will never be triggered. Although not required to look for gaps, node A still needs to register the tracer reports, because in the early runs they are the only evidence that the tracer waves are running.

As mentioned, in approximately the same time period where node A acts as base node and runs tracer waves to node B, a similar sequence of events takes place, running on oppositely directed cells. Node B sends out *tracer init* volleys toward node A, monitors the *tracer reports* and *arrival volleys* it receives, then concludes that it is on the same contour as node A.

Once it is established that nodes A and B are on the same contour, it is desirable to "save" the fact, as in a computer, in other words record it as some sort of a hardware change that can later be "read out."

In this model we choose the form of "recording" to be the creation of a *link* between the two nodes, via fast axons, with an arrangement whereby ignition in either one of the nodes causes ignition in the other (and reverberation is prevented by the refractory period). The axons, and the synaptic changes which enable them to pass ignition between the bridgeheads, make up the memory hardware in which the same-contour relation is "saved."

The "read-out" of the same-contour information is done by the subsequent co-ignitions, and as part of the message contained in them. The fact of simultaneity says that the nodes are on the same contour.

A separate part of the message is conveyed through the neuron type making up the bridgeheads. When the bridgeheads are made up of *contour cells* (as in this chapter), they convey contour orientation, color contrast, and visibility information;

when they are made up of *direction-coded cells* (as in Chapter 20), they convey relative direction information.

19.1.12 Node A Knows that the Reaching Is Bidirectional; So Does Node B

Returning to the sequence of events in the linkup, and noting that node A now knows that it is on the same contour as node B, we next note that node A also knows that there is *two-way reaching* between it and node B.

Node A knows of the two-way reaching by repeatedly detecting the way its *tracer init* volley is followed by an incoming tracer shower continuous with the *arrival volley* from node B. It knows that it reaches node B because otherwise it could not have initialized node B to send the *arrival volley*. It also knows that it is reached by node B, because otherwise the *arrival volley* could not reach it. And in both cases it knows that its own ignitions cause the sequence of events it detects, because they are initiated on a random schedule under its own control.

The two parts of the test, in other words, do more than just tell us, human observers of the situation, that A and B reach one another; they also see to it that node A *knows* of both halves of the bidirectional reaching. Incidentally, this is consistent with what was seen in Fig. 16.1 that a bidirectional active link enables neurons in a node even to detect when the linkage, in either direction, is *getting weak*.

19.1.13 Linkup and Active Link Operation

It still remains to interconnect the two nodes through an *active link*. To do this, the same axons which formerly carried the *tracer init* volley from node A to node B will provide one lane of the pathway between nodes, and the axons which formerly carried back the *tracer report* from node B will provide the other lane.

During *tracer wave operation* the self-ignitions of node A are interpreted as *tracer init* volleys, which do not cause immediate response, only response upon arrival of the next contour wave. In *active link operation* identical-looking volleys are interpreted as "*echolocation volleys*," which cause immediate response by node B ("*echo volleys*"), and no response from the cells in-between.

In the active link operation, nodes A and B both continue self-ignitions; the *echo volleys* of each node sometimes divert the node's self-ignitions (by starting their usual timer after the *echo volley*), and when the self-ignitions come right before an echolocation volley, pre-empt the echo volley. The result is that the two nodes sometimes co-ignite and at other times ignite alone. As mentioned before, this does not disrupt the message of the co-ignitions, since those are still much more frequent than the accidental rate, and therefore enough to broadcast that the two nodes are on the same contour.

Each co-ignition after contour linkup is restricted to the contour *bridgehead* cells of its node. Since the two contour bridgeheads of a node are made up of non-overlapping sets of cells, co-ignition of a node with one neighbor does not interfere with co-ignition of the same node with the other.

19.2 Operating Modes of Neurons

When looking at the foregoing description of contour linkup, one is justified in asking, how the cells in nodes receiving the *tracer init* volleys, which call for delayed response, manage to distinguish those volleys from the *echolocation volleys* of the active link operation, which call for immediate response. Since the ignitions are faceless, their form does not offer any way to detect a difference.

The issue (which recurs throughout the monograph) presents a problem which, as far as I can see, cannot be solved without making a drastic change in our picture of the neuron, and postulating that the neuron is *capable of different operating modes*, in which its actions are different. The interactions between its synapse pools are different and the volleys arriving to the synapse pools call for different responses.

It may be suggested that the drastic conclusion may be avoided by saying that instead of one set of neurons moving through different modes, different sets of neurons could be envisioned, each with properties suitable for a different mode of operation. In the case where *tracer action* changes over into *echolocation action*, one should maybe envision one set of neurons in node B programmed to respond to volleys from node A by setting itself up to ignite then, detecting the next contour wave, and another set programmed to respond to the same volleys by igniting immediately, and then one should envision switching from the first set to the second at some point.

However, the problem is that the second set would have to rely on synapses marked out during the first phase of the operation, and synapse marking is not transferable between neuron sets. Synapses in node A marked out by volleys emitted by one set of node B neurons will be missed by volleys coming from another set, and the synapses marked out on the node B neurons themselves also cannot simply be transplanted to other neurons.

In fact, the transfer of marked synapse sets is a recurring theme in mode transitions. In the sequences of drawings in the next few chapters, a typical mode transition is where some neurons receive repeated input on one synapse pool, and as soon as they decide that the input is unambiguous, and mark the synapses, they move to the next mode, where they use the input to the newly marked synapses to control the processing of input arriving on some other synapse pool.

The conclusion is that the neurons must detect the mode they are in, and act accordingly. The ability to recognize operating mode is then the answer to the problem of distinguishing between indistinguishable volleys like the *tracer init volley* and the *echolocation volley*. The interpretation of the volleys is inherent in the *mode* of the cells during the times when the volleys arrive, or equivalently inherent in the *phase* of overall operation.

One of the challenges in the coming chapters is to demonstrate that different neurons, in deciding the mode they should be in, are not led to draw inconsistent conclusions from the signals arriving to them. An added problem is that the tests by which modes and phases of operation are recognized are not necessarily simple, since they often involve looking for specific features in the behavior of multiple signals on interacting synapse pools (Figs. 16.1(c–e), 17.1(h)).

Accordingly, it is hard to envision the *statistical engine* and the *mode control mechanism* in neurons without invoking molecular signaling mechanisms of some form. The microtubules and microfilaments in the neurons are generally not believed to be suitable for the role of an information transfer network, instead, they are believed to be part of the substrate of cell growth (Tomasek and Hay, 1984). On the other hand, the movement of molecules through the intracellular fluid can be fast enough to bridge distances on a time-scale matching the surprise detection and mode transitions, also, as far as is known, molecular signaling systems can be envisioned as carrying the stamps of specific synapse pools, making decisions, and effecting changes.

19.3 Firing Games: Goal-Directed Organization Without a Leader

The operating modes, together with the rules of recognizing them, once implemented, can accomplish a feat which may at first glance appear impossible: coordinated goal-directed action in which many neurons participate, and among the neurons none plays the role of a leader.

What makes this possible is that all the interacting cells and localities share the same preprogrammed *mode* definitions. The shared rules are what lend the overall activity the appearance of coordination. The challenge is to invent a method whereby all localities involved in a task, in their own private ways, recognize the global situation as requiring the same mode; arrange transition to the mode; and the processing is not disrupted by momentary differences among individual cells during the intervals of transition between modes.

The actions taken during modes are chosen from a limited menu: creating ignition-support contacts; ignition; starting self-ignitions; stopping the self-ignitions; marking a set of synapses; setting up passive following of volleys to marked synapses; marking the synapses from another group passively responding to the local group.

The cues for transition to a mode are along the lines seen in Fig. 16.1 and elsewhere, invariably subject to the requirement that sufficient surprise be imparted to the cells to identify the required mode with certainty.

The pathways along which mode transition cues travel have to be assembled before they are to be traversed, while in the network as a whole many processes go on independently without disturbing one another.

Every mode, in addition to synapse modification rules and firing rules, is characterized by some sort of a goal, whose achievement marks the end of a phase and transition to the next one. At each location the network design must set up a sequence of phases properly supporting the next phase in other localities.

One problem with using *quantity of surprise* as the criterion for achieving the goals is that the accumulation of surprise contains an element of graduality. Mode transition is a qualitative change and there is nothing gradual about it, therefore it requires a *threshold* to be applied against the surprise. But that means that different locations will reach their thresholds at different times.

Accordingly, as mentioned, the over-all network design must be robust enough to be free of "race conditions," situations which would make the overall outcome sensitive to the order in which different places reach their thresholds.

The leaderless multi-mode actions designed along the lines just described will be called "firing games," to suggest the similarity to some *team sports* where a team can smoothly fight its way to winning a game without anybody issuing any orders.

The rest of this book will only contain a few examples of firing games: the contour linkup (in this chapter), the chain linkup, the triangle closing, and the all-to-all linkup. This is clearly a tiny sample of all the tasks that must be addressed by the visual cortex, but I think, maybe, it will be enough to illustrate the general methods that may be used in designing such protocols.

Let me next return to the contour linkup whose reasoning was sketched out in Sect. 19.1.

19.4 Directional Specificity of Contour Cells: R-cells and L-cells

Contour cells have in common with direction-coded cells the general *labeled lines* coding philosophy by which they convey spatial relations to the upstairs; in both cell types, "*the messenger is the message*." Different values of a parameter are sent out by different cell pools, and each pool is, by and large, dedicated to a value of the parameter.

The difference is that direction-coded cells must distinguish between as many directions as is allowed by their angular range, while contour cells only need to distinguish between two directions. They must either act only on the cells to the right of them on the contour, or only on the cells to the left of them.

For reference, I will call the two directionally specific cell types R-cells (right-linking cells) and L-cells (left-linking cells), and denote them in drawings by oppositely oriented half-ellipse designs (see for instance Fig. 19.1). Pairs of L-cells and R-cells so situated that they can participate in opposing bridgeheads on interlinked nodes will be referred to as "facing conjugate" cells. R-cells and L-cells can send local orientation and color opponency information to the upstairs, and it will be noted that facing conjugate cells are identical in regard to these properties.

19.5 A Note on the Drawings Describing Contour Linkup

The remainder of this Chapter is organized around several series of drawings, meant to sketch out the contour linkup process from a number of different points of view. Fig. 19.1 is a brief summary, leaving out many details which are then added in

Figs. 19.2 and 19.4. The two latter sequences approach some of the same things from different perspectives, each of them revealing some aspects of the process the other cannot. By necessity, there is some redundancy among the drawings, for which I apologize.

In Figs. 19.1(a)–(f), 19.2(a)–(j) and 19.3(a)–(d), the *tracer init* volleys, and the tracer waves they initiate, are drawn as always moving from left to right; in Fig. 19.4(a)–(l) both directions are included. The direction-limited drawings serve to emphasize the difference between the roles of the two kinds of contour cells in a tracer wave system. When one kind sends the *tracer init* volleys, the other kind has to send the *tracer reports*. The two kinds, it will be noted, merely correspond to different operating modes of contour cells.

It will be noted that the icons representing R-cells and L-cells contain separate synapse sets at their *contour wave input* end reserved for the two propagating directions (dromes) of contour waves, indicated by separate semicircular marks at the bottom side of the half-ellipses (see for instance Fig. 19.2(a)). The separate synapse sets enable the contour cells to respond to each drome of contour waves in accordance with their operating mode. The result is that the processes moving in opposite directions do not interfere.

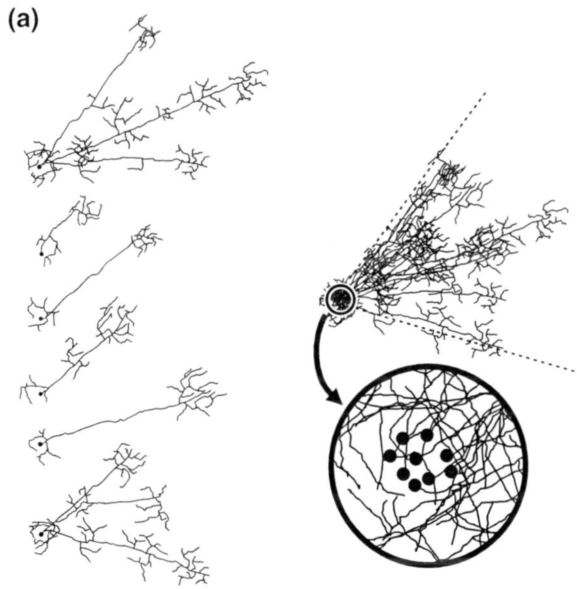

Fig. 19.1(a) Elements of the contour linkup. Some of the right-linking (R-type) contour cells in a node. Some R-type contour cells which may exist in a node are sketched individually (*left*) and superimposed (*right*). The long straight sections of the axons (generally longer than shown) are Gennari fibers or U-fibers. The cell bodies are shown stylized as *small circles*. The recurrent collaterals, arborizing in the vicinity of the cell bodies, support the ignitions of the R-cell bridgehead, some shown magnified in the inset. (Hand drawings, not to be taken as data)

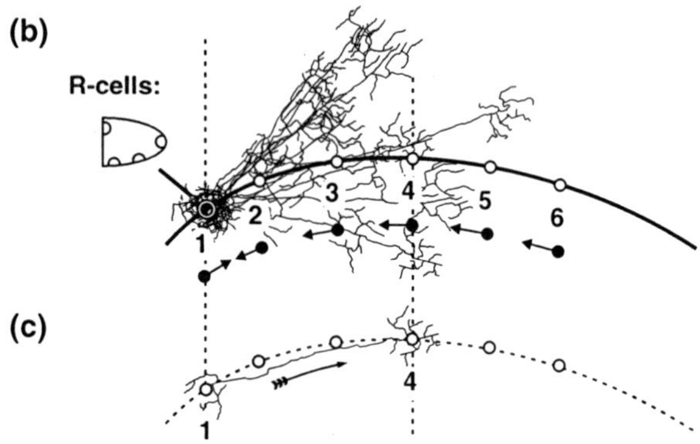

Fig. 19.1(b)–(c) Elements of the contour linkup (Cont.). R-type volley, meant to reach all L-cells to its right. Each of the numbers 1, 2, ... 6 refers to a set of contour cells. Of these, 1 represents a node; the rest are neurons which individually stand by when detecting the presence of contour waves at their location. The volley initiating tracer action (tracer init) is sent out from a set of R-type cells at location 1, on all the axons shown, and aims to reach all locations 2, 3, ..., up to the next node. The half-ellipse design shown in the inset is used here and in the subsequent figures to represent R-cells (with a similar design for L-cells, see (d)). Drawing (c) shows a typical one of the axons participating in the ignition and reaching location 4 (axonal drome indicated by "Cajal-style" arrow)

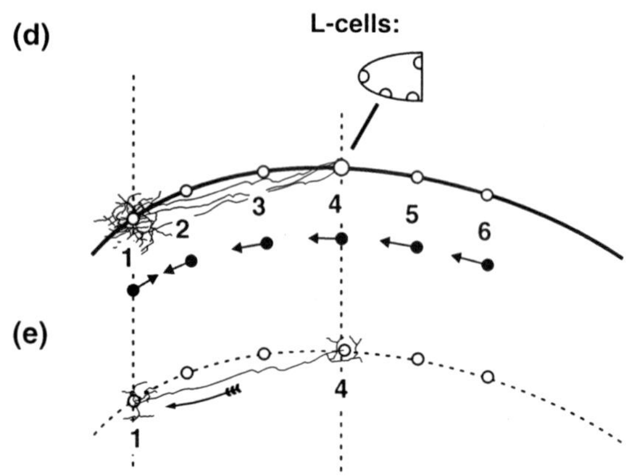

Fig. 19.1(d)–(e) Elements of the contour linkup (Cont.). The volley only reaches some of L-cells at any location. When the initialized L-cells, a little later, send out a tracer report, the reports go out toward the left. Only the subset of L-cells reached by the tracer init volley sends out tracer reports, accordingly, the reports tend to be approximately aimed at node 1, because in many L-cells the connectivity between locations is expected to be roughly reciprocal. (**e**) shows a small effective field L-cell reaching some neurons at 1 from location 4

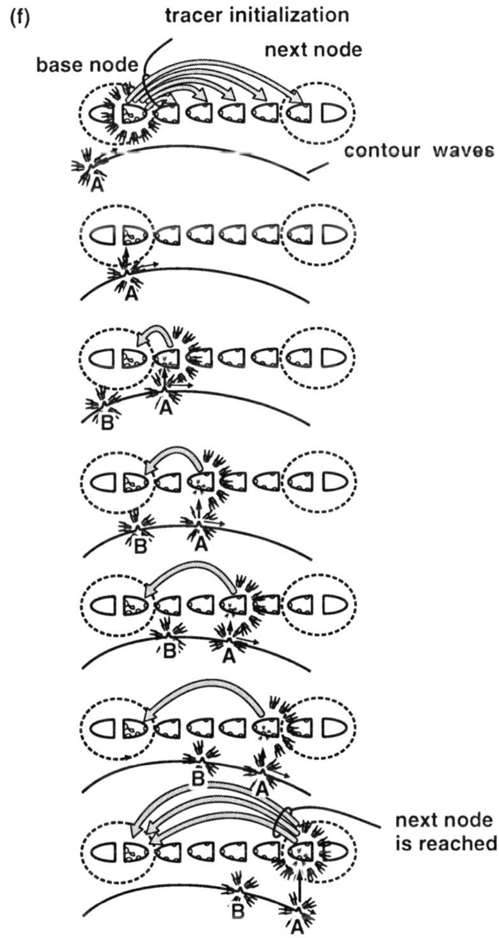

Fig. 19.1(f) Elements of the contour linkup (Cont.). Summary illustration of a tracer wave run. This sequence summarizes the main steps, leaving some of the details to the next sequences (Figs. 19.2, 19.3, 19.4). Ignition of the base node (*top drawing*) reaches all L-cells to its right as well as the L-cells of the next node; it tells them to stand by until the next contour wave comes, respond when it does, then reset itself. When the next contour wave (marked "A") passes "beneath" them, the in-between cells fire individually (next series of drawings) and the next node ignites (*bottom drawing*), giving off a recognizable volley. Since the cells are "reset" after giving off a contour wave, until initialized again, a second contour wave (like the one marked "B") has no effect. All these responses are sent off by long-axon neurons and reach the base node (*arrows pointing left*), and they together convey enough information to tell the base node whether an uninterrupted contour connects it with the next node. The drawing ignores a number of items taken up in the next sections, for instance the adaptive achievement of reaching, the need for marking the synapse sets prior to using them, and the "second enable" mechanism

Several of the cells are drawn as having "clustered" axonal arborization, and two as having long axonal branches extending in several directions. The latter cells cannot be direction-coded since in them the direction of the axonal arborization is indeterminate, but direction coding is not important in the contour cells, only direction asymmetry is, the kind needed to distinguish between the two directions along a contour.

The *tracer init* volley should more properly be described along the lines of the "omnidirectional" volleys (Sects. 20.1.4–20.1.10) created by purposefully making a set of original volleys spread to many differently aimed cell groups at first, then dropping most of the added groups after identifiable responses are registered, but in this section there is so much complexity from other sources that we will stay with the simpler description. Also, it is possible that along with the R-type and L-type contour cells, the *direction-coded* cells participate in the tracer dialog, because their reciprocal connectivity is expected to be more reliable and therefore their use saves resources, but this concept, too, adds complexity, and it will not be pursued here.

19.6 Synaptic Interactions During Tracer Runs and Linkup

The description of tracer runs is incomplete without an attempt to spell out the different information streams arriving to the different synapse sets of the participating neurons. The first of these is the contour wave input, coming from the complex and hypercomplex cells; it is what tells the neurons that a contour wave is present and they are called upon to do something. It will be noted that it is necessary to envision distinct synapse sets for the two different dromes of contour waves (Fig. 19.2a), because only one of these is used by any group in triggering its tracer wave output. Other synapse sets distinguish between local and remote inputs, others again between inputs from contour cells similarly or oppositely arborizing.

The synapse groups in (a) are numbered as follows: 1, 2 – Inputs from the underlying contour waves with separate synapse sets serving the two directions of contour wave movement. 3 – Gennari input. *Tracer init volleys*, *tracer reports*, and *arrival volleys*, all arrive on these synapses, as do the later *echolocation volleys* and *echo volleys*. The synapse sets bringing self-ignition volleys from other nodes are marked earlier, and could have been drawn as being separate from the ones bringing the tracer reports, but were combined to reduce clutter. 4 – Input from neighboring L-cells used as the "second enable" for tracer action (the "first enable" being the *tracer init* volley), making sure that all tracer waves start at the node and none can start in mid-interval. The first few tracer runs ignore the second enable requirement to permit marking its synapse set. 5 – Local ignition support synapses in the base node.

The continuity detection, in terms of spikes received and sent out by the R-cells in the base node, is sketched in the drawings in Fig. 19.3. As seen in Fig. 19.3(b), the input pattern sought is one of uninterrupted bombardment starting with arrival of the first contour wave after tracer init, and ending with the arrival volley from the node at the right end.

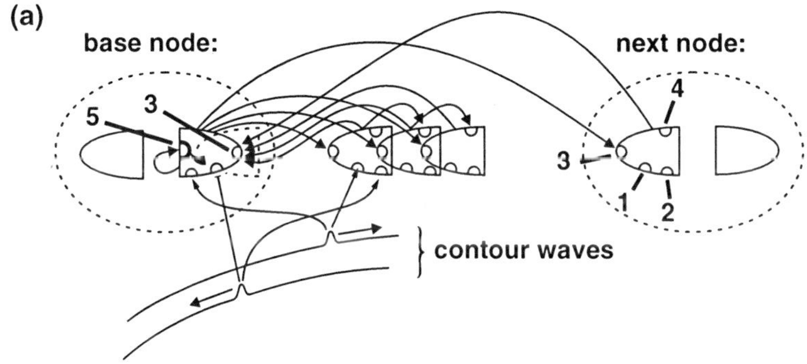

Fig. 19.2(a) Continuity detection by use of tracer waves. Notation: summary of the synapse sets. Two nodes are shown (*circled at left and right*) and some L-cells in-between. Each node contains a set of R-cells and a set of L-cells, which are "local conjugates" (their axonal arborages leave at opposite poles of the node). Among the cells in-between only those are shown which are L-cells and facing conjugates to the R-cells of the node at the left (their axonal arborage points toward them). A set of in-between cells is included, and shown as being placed close enough together that each is reached by the short-range contacts from the one at its left

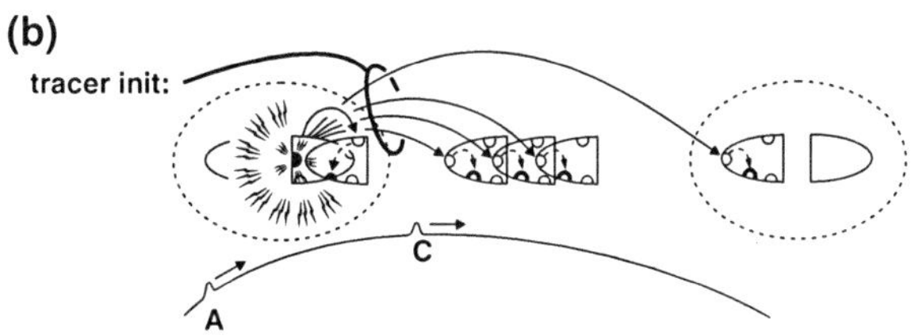

Fig. 19.2(b) Continuity detection by use of tracer waves (Cont.). Tracer wave initialization (tracer init) volley. Self-ignition of R-cells in the node at left sends volleys to all L-cells to the right of the node, including the L-cells in the node at right. Arrows inside the neuron sets show the cross-potentiation from the tracer init to synapse set 1, receiving contour wave input (see also Fig. 19.3). The first few L-cells in the immediate vicinity of the igniting node, become fully initialized when reached by a volley through short local axons (they do not require a "second enable"), to respond when the first contour wave reaches them (synapse set 1 blackened). The effect of the tracer init volley on all the other L-cells is to supply the "first enable" input (synapse set 1 drawn in heavier line). The tracer initialization only lasts until firing (see (d)), after which it is undone until the next tracer init volley

The *tracer init* volleys occur repeatedly but the first few have no effect because the recipient neurons must first mark the synapses on which they arrive. Regardless whether a volley succeeds in eliciting a useful tracer response, the cells emitting the tracer init must set themselves up to receive the responses after each self ignition (Fig. 19.3(a)).

It is noted that the *tracer init* volleys are autonomously generated at the base node, on a random schedule, and that this schedule is independent of the random schedule on which the underlying contour waves arrive. Therefore in general there will be a brief time interval between the *tracer init* volley and the arrival of the next

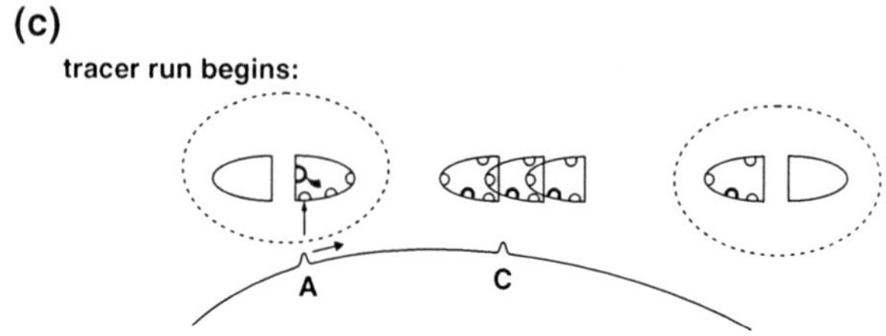

Fig. 19.2(c) Continuity detection by use of tracer waves (Cont.). First contour wave after the volley starts the tracer run. Continuity detection begins when the first contour wave after the tracer init reaches the base node. The contour wave in question is denoted "A" in these drawings. The progress of two other contour waves is followed in the drawings: "C" comes too early, and is already past the base node when the tracer init is emitted, and "B" (see (e)) comes too late. The system of "second enable" prevents C from triggering tracer reports

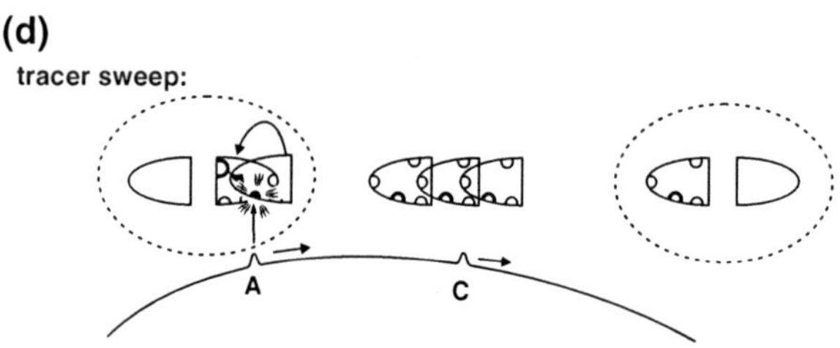

Fig. 19.2(d) Continuity detection by use of tracer waves (Cont.). Tracer wave starts at the first cells after base node. The first cells to the right of the base node, within reach of the short axon collaterals, fire as soon as the contour wave "A" reaches them, and send tracer reports to the base node as well as "second enable" signals to the cells at their right (not shown)

contour wave, the one to be traced as a result of the volley. This is why the continuity detection is shown to begin, not at the moment of the *tracer init*, but at the slightly later moment when the contour wave "A" passes through the node.

The contour wave "B" in (e) passes the base node, but has no effect, because the tracer initialization is only valid for one tracer report, and its effect disappears after the tracer report, and is only revived at the next tracer initialization. Similarly, the contour wave "C" has no effect on any of the cells it passes because of the *second enable* requirement. The *second enable* signal, emitted a moment earlier when "A" passed the base node, only reached the cells to its immediate right, and the cells around "C" were too far away. Since the distance between "A" and "C" remains constant, the cells over "C" are never enabled.

The arrangement permits the base node to monitor one tracer wave at a time, along the lines of Fig. 19.3. Also, the arrangement relieves the base node of any time pressure; the node can take all the time it requires to complete its analysis of each tracer run, because it has control of the timing of *tracer init* volleys. It is free to delay emission of the next volley (in other words delay the charge leakage into the bridgeheads, as in Fig. 7.1) until it is finished analyzing the tracer wave in progress.

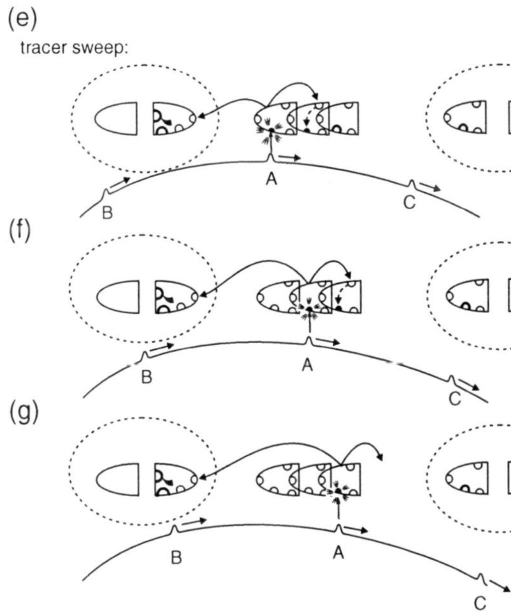

Fig. 19.2(e)–(g) Continuity detection by use of tracer waves (Cont.). Tracer wave moves through successive cells between nodes. After leaving the start zone, the tracer wave, wherever it is, contributes to the initialization (second enable) of the cells immediately to the right of its wavefront. In (**e**) the contour wave "A" fires the first of the three cells shown as being near each other (initialized by other cells not shown). The firing sends a tracer report back to the base node via Gennari axons and a "second enable" pulse to the cells at the immediate right initializing them (blackened synapse set). In (**f**) the contour wave reaches the cells enabled in (e), fires them and again sends a tracer report to the base node and a second enable to the right. In (**g**) the same happens at the next cells

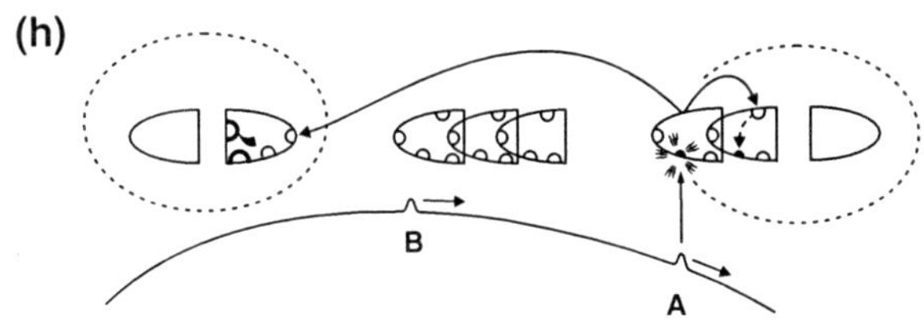

Fig. 19.2(h) Continuity detection by use of tracer waves (Cont.). Tracer wave nears next node and enables contour wave response. When the tracer wave makes its way to the cells right before the next node, they initialize the first cells of the node in the same way as they would if they were individual cells as in (e)–(g), rather than members of a node

It will be noted that the *second enable* machinery only works because neurons can distinguish between inputs coming from near and inputs coming from far.

The "arrival volley" is distinguishable from the "tracer reports," because the synapses on which it arrives are already marked in the base node. When the arrival

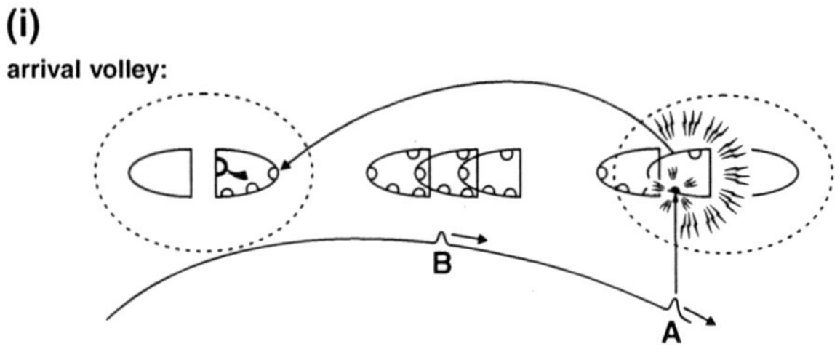

Fig. 19.2(i) Continuity detection by use of tracer waves (Cont.). Ignition of next node (arrival volley) indicates continuity. When the contour wave "A" reaches the first cells of the node enabled in (h), the cells fire; then because of the ignition support contacts, the firing spreads to the rest of the ignitable group, igniting it (see larger circle of "sparks")

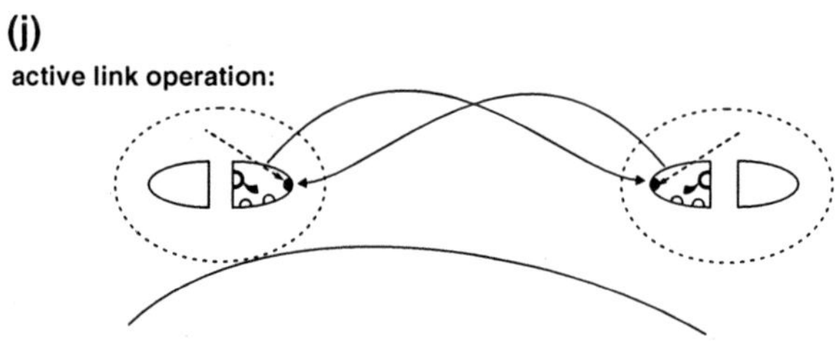

Fig. 19.2(j) Continuity detection by use of tracer waves (Cont.). New operating mode: repeated co-ignitions. After sufficient repetition of the tracer run and arrival volley, the base node knows (with a fair degree of certainty) that the next node is on the same contour as it is itself. It also knows that its ignitable group of R-cells reaches the next node, otherwise the tracer init would not have been able to set it up for arrival response. It also knows that the group is reached by the next node, otherwise it would not be reached by the arrival volley. In other words, the base node knows that there is bidirectional reaching between it and the next node

volley is registered at the base, the pattern of tracer waves continuously leading to an arrival volley (Fig. 19.3(b)) provides the base node with confirmation of the continuity.

Independently, the next node also goes through the same continuity detection, in the opposite direction, with a base node made up of L-cells and with tracer reports and arrival volleys given off by facing R-cells (as seen in Fig. 19.4). The need for this opposite-directed process exists because at the end of Fig. 19.2 only the left-hand node (the base node of Fig. 19.2) knows of the bidirectional reaching whereas the right-hand node does not, until it performs continuity detection of its own.

Because the objective facts remain the same whether tested from one node or from the other, the results of the two kinds or runs will also be the same, and the next node will, sooner or later, also undergo the same mode transition.

Having established bidirectional reaching as well as continuity, the two nodes are now ready to form an active link, enabling them to perform adaptive maintenance of the contacts and neuron sets in the face of contour drift (Chapters 17, 18).

In the new mode, the underlying contour waves are ignored. The cells of the bridgeheads will continue their self-ignitions, while they passively follow the self-ignitions coming from the next node. The self-ignitions in the new operating mode are renamed *echolocation volleys* and the responses *echo volleys*, to underline their different natures. The self-ignitions at the two ends will run on different schedules, independent except to the extent that refractoriness causes the two processes to keep out of each other's way. Refractoriness, also, prevents the ignitions from bouncing back and forth between the nodes.

19.7 Continuity Detection from the Standpoint of the Base Cells

The next drawings show the clues received by the base cells used in making or not making a mode transition during the process of continuity detection.

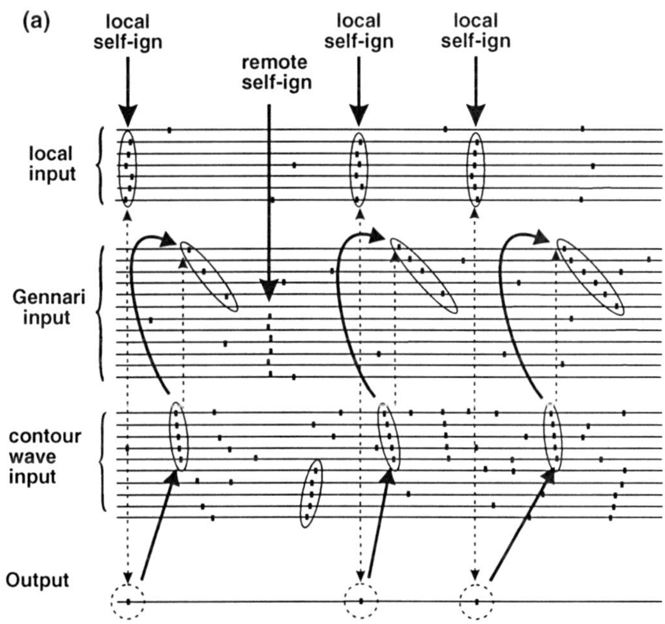

Fig. 19.3(a) Continuity detection from the standpoint of the base cells. Events during the warm-up runs of tracer setup. Inputs and outputs of a typical neuron of the base node ("test neuron") during contour linkup. The output spikes of the test neuron are shown at the bottom. Three tracer init volleys (circled in "local input" traces) are shown as they are generated in self-ignitions, along with the spike emitted by the test neuron, which participates in the ignitions ("output" *trace at bottom*). Broken lines show that the emitted spikes are part of the activity giving rise to the tracer init volleys. In the initial runs, the stage shown here, the ignitions of nodes (*arrows on top*) are still in the stage of being recognized as tracer inits, and their synapses are still in the process of being marked; further, the tracer waves are not yet subject to the "second enable" input requirement, in order to give their synapses a chance to be marked (Sect. 19.1.8)

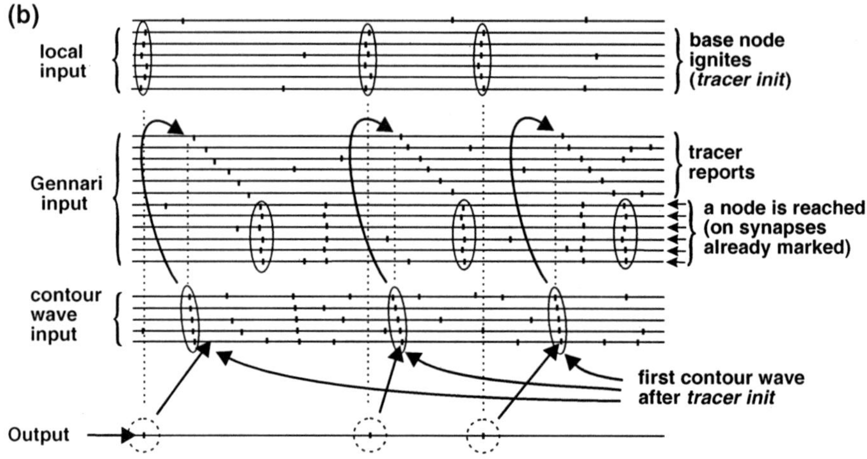

Fig. 19.3(b) Continuity detection from the standpoint of the base cells (Cont.). Events when the outgoing tracer runs are successful. In this drawing each locally generated tracer initialization volley (*circled*), upon arrival of the next contour wave (*circled*) is reproducibly followed first by a run of tracer reports on the "Gennari input" traces and then immediately afterwards, also on the "Gennari input" traces, by an arrival volley (*circled*). The arrival volleys come in on synapses which occasionally also bring other remotely generated volleys (not circled). The second enable action (not shown) is now fully operational

In the initial runs shown in this drawing the tracer runs have gaps. The next node, being far away, is not yet reached by the *tracer init*, and there is not yet any sign of an arrival volley. Since the second enable requirement is not yet imposed, all Gennari neurons in the traces which are not part of nodes are fully enabled as soon as they receive a tracer init, and fire when they detect the next contour wave.

The "contour wave input" traces, from the local complex/hypercomplex cells, indicate the first contour wave arriving after each *tracer init* (circled), along with one extra contour wave included for illustration (not circled), which is ignored because it is not the first one after a *tracer init* (like wave "B" in Fig. 19.2).

The neurons emitting the tracer reports can hear the same contour wave heard by the local neurons, at the somewhat later time when it arrives to them, as seen by the staggered spikes generated as the successive cells are triggered when the contour wave makes its way to them.

The contour waves associated with the opposing *tracer init* volleys travel in opposite directions and on different sets of neurons. Their tracks could have been separately labeled, as in Fig. 19.2; instead this drawing simply puts their spikes on different tracks, and "slants" the two kinds of volleys oppositely. It is seen that the backward track of tracer setup is also in its initial stages, otherwise the next "backward" contour wave after the remote self-ignition would cause a local ignition.

If after some time still no response is received on these marked synapses, the cell (and others like it) will conclude that the bursts A1, A2, and A3 come from another contour (Fig. 19.6), and accordingly their "tracer init" effect on the local node is cancelled.

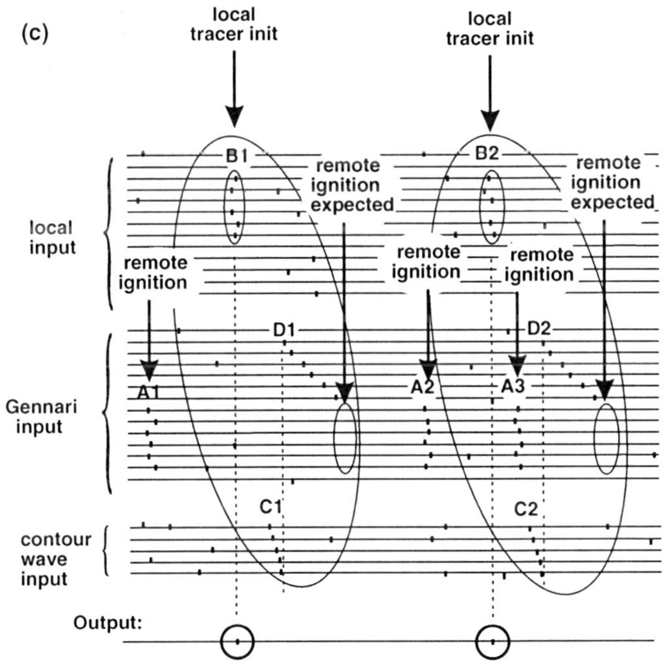

Fig. 19.3(c) Continuity detection from the standpoint of the base cells (Cont.). Events when outgoing tracers fail to locate a node. In this example the local node records the synapses on which remotely generated volleys (A1, A2, A3) arrive, and on which the "arrival volleys" are expected to arrive when they do. However, after local tracer initialization volleys (B1, B2) are issued, and are, upon the next contour waves (C1, C2), as expected, followed by tracer reports (D1, D2), there is no sign of any remote volleys on the marked synapses, at the times when the "arrival volleys" are expected

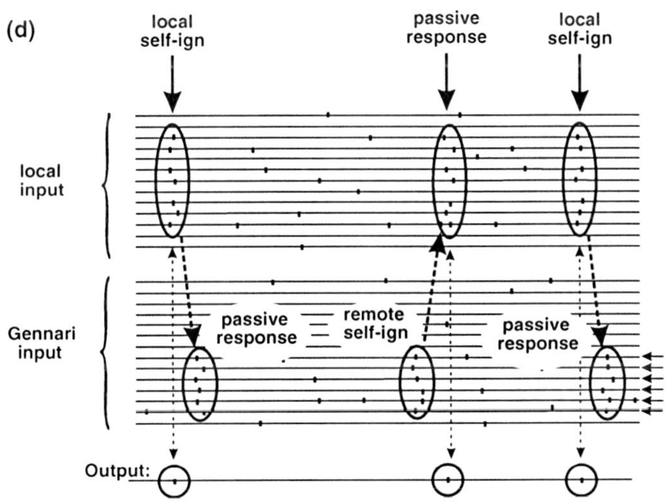

Fig. 19.3(d) Continuity detection from the standpoint of the base cells (Cont.). Events after successfully completed linkup. This drawing shows the sequence of events after the mode transition which changes the interpretation of ignitions at the base node from tracer init volley to echolocation volley, and the ignitions at the other node from arrival volley to echo volley

After the sequence shown in (b) has been repeated enough times to establish linkage, the rule by which the neuron responds is changed. Ignitions from the remote node are now immediately followed by local ignition in passive response to the volley arriving to the marked synapses (arrows at right), and without the intermediate step of waiting for the next contour wave.

Since sooner or later the remote node similarly drops the step involving contour waves, the local self-ignitions soon (or already earlier) begin to be followed by passive responses by the remote node (synapses are marked at right). Because the previous runs have confirmed one other, consistency is maintained and the set of remote cells responding to local volleys is the same as the set to which the local neurons respond.

19.8 "Understudy Processing" of Cells Before They Join a Node

As before, an extra layer of complexity is added to the linkup protocols when it is recognized that the retinal image drifts, and new neurons keep having to be added to the node as old ones are dropped.

As was seen, the cells of each node go through a sequence steps, and in different steps the volleys arriving to the cells are differently interpreted. The problem is that a neuron just entering the node cannot have any memory of earlier phases of the linkup, since those were experienced by other neurons; the earlier state of the newly entering neuron was one where it was outside the contour altogether, and was, by and large, "not doing anything."

However, the problem can be solved by making use of the fact that the signals getting to the neurons near a node are not much different from the signals getting to the neurons participating in the node (Fig. 18.1). Once one accepts that neurons and neuronal localities go through "modes" of operation, it becomes clear that the same cells which regulate the modes of neurons already participating in the node can also act on the cells nearby, the cells which are in effect the "understudies" of the cells in the node, and paint them with the same modes.

The only difference between node members and the understudy cells in the process of being "warmed up" is that the latter are not allowed to emit spikes until they receive contour wave input.

But the understudy cells can hear the Gennari volleys, the tracer showers coming from the Gennari fibers, and the volleys emitted by the local neurons. Accordingly, they can mark the synapses through which each of these inputs will later arrive when (and if) they get to join the node. The only thing left for them to do when the contour waves move into range (telling them to join the node) is to fire in imitation of the cells in the node.

Subsequently the other node cells, which have not previously heard them fire, mark the synapses from the newly joining cells and add them to the synapse set of node ignitions.

The length of time a neuron spends as part of a contour string or as part of a node is determined by the rate of image drift and, in the case of nodes, by the node size. It is often much shorter than the time it takes for the node to progress through all phases of a linkup.

Each phase of a multi-step linkup process, for instance the phase of tracer runs in the contour linkup (Fig. 19.4), typically requires multiple repetitions of volleys, which means that the phases progress in their own time frame. They cannot be hurried up to match the rate of image drift. Therefore, when the image drift is fast, a cell sometimes cannot even finish a single phase before it must leave the node, and at other times it only has enough time to end one phase and begin the next one. This is probably enough time, but when the image drifts too fast to permit understudy processing, tracking will fail.

19.9 Phases of a Linkup, with Each Phase "hammered in" by Repetitions

The next series of drawings (Fig. 19.4) shows, in heuristic terms, the cross-currents of information coming to cells from two different directions, as linkup initiated from opposite ends of a segment of contour give rise to tracer runs in both directions.

The similar appearance of volleys carrying different messages in different contexts must be reconciled with the requirement to "hammer in" the messages through repetition. It is not immediately obvious that this can be done at all, since the "hammering," which calls for repeating certain events a number of times not known a priori, appears to be incompatible with interpretation of the volleys through a pre-

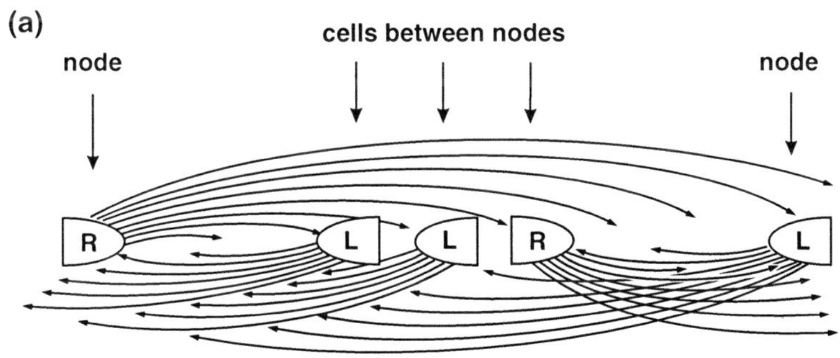

Fig. 19.4(a) Implied messages during bidirectional contour linkup. Notation. Two nodes and the interval between them are shown. Only the contour cells are of interest. In the nodes only the cells with axons pointing toward the interval are shown (R-cells at *left* and L-cells at *right*) and in the interval between nodes a few examples of the cells are shown which are not members of nodes but mediate continuity detection. The latter cells act as individuals, not as ignitable groups, and there are enough of them that between them they reach all nodes to their right or left which they need to reach

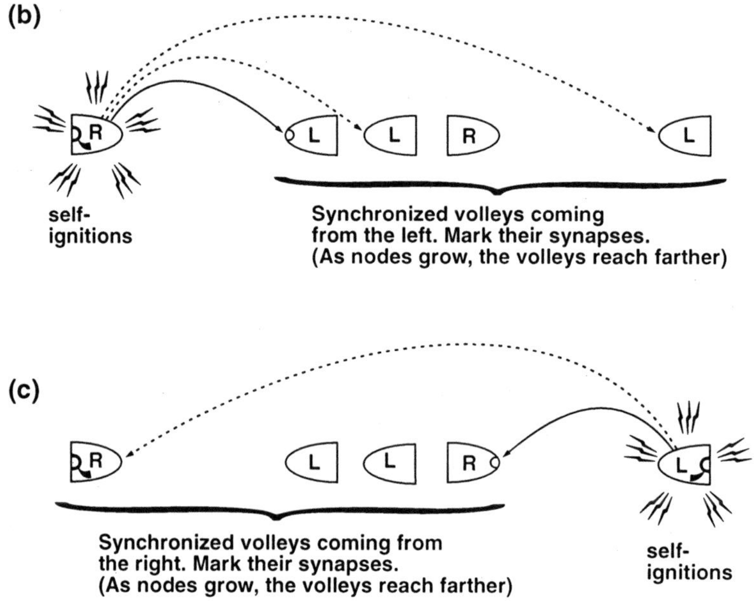

Fig. 19.4(b)–(c) Implied messages during bidirectional contour linkup (Cont.). Nodes start self-ignitions and extend their reach. At first, nodes are not ignitable groups but only send out random firing. After a while they become ignitable and emit self-ignitions on a random schedule. When they do, the cells reached by their volleys can mark the synapses on which the volleys arrive

arranged sequential order, such as toggle action. (The manner in which the requirements may be reconciled is illustrated in Fig. 19.3.)

At the beginning of a period of fixation the nodes are not yet fully formed. The contour cells in the nodes at the left and right are in the process of becoming ignitable groups. In the process of forming ignitable groups, group members must first mark the synapses from other group members (ignition support links). From the standpoint of cells outside the nascent nodes, the firing in this initial stage appears random, lacking synchrony and surprise.

A little later, having developed their ignition support synapses, the cells of interest, R-cells in the left node and L-cells in the right node, proceed to move to the next phase, in which they repeatedly perform self-ignitions (crescendo sign). From the standpoint of the cells reached by them, the ignitions are now surprising, and each repetition increases the surprise. The cells reached become aware that they are to play a role in a linkup when they detect the underlying contour waves, telling them that there is a contour in their receptive fields.

The ignitions from the nodes may not right away reach the contour cells all the way to the next node; the drawing indicates this by showing some of the longer contacts in broken lines. In time, as the nodes grow sufficiently, the reaching is extended (Chapter 16). Wherever the cells are reached, the synapse sets of the volleys are marked (semicircles).

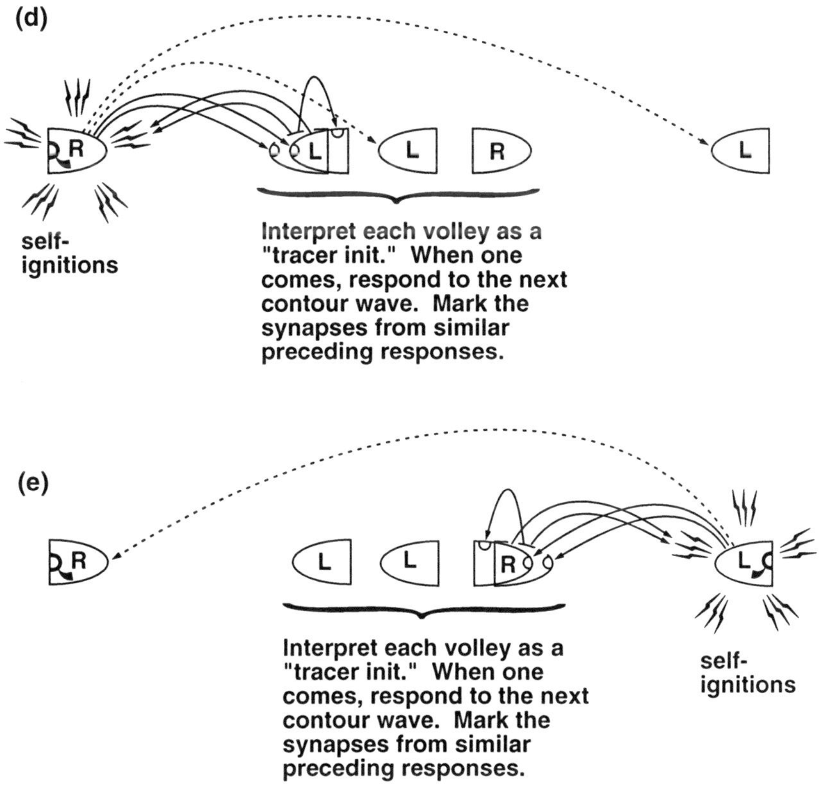

Fig. 19.4(d)–(e) Implied messages during bidirectional contour linkup (Cont.). Cells reached by ignitions run tentative tracer waves. Volleys coming from facing conjugate cells are interpreted as tracer initialization volleys (for L-cells the volleys from R-cells, and for R-cells the volleys from L-cells are interpreted as tracer inits). After a few repetitions, the cells reached by the volleys respond to the next contour wave after each volley. Meanwhile the cells mark the synapses from neighboring like-facing cells emitting tracer reports before them for subsequent use as "second enable" input terminals

As each cell is reached by the ignitions (coming from its facing direction), it moves to a new mode, the *tracer warmup* mode, where it interprets the ignitions as tracer initialization (tracer init) volleys, starts paying attention to the underlying contour waves (Fig. 19.2), marks their synapses, and emits a spike (tracer report) when the first contour wave after each volley arrives. It ignores all subsequent contour waves until the next tracer init volley.

A separate process in this phase is marking input terminals from neighboring contour cells of like axonal orientation (L-cells mark the output from L-cells at their immediate left and R-cells mark the output from R-cells at their immediate right), provided they emit tracer report spikes right before the cells which do the marking. This process marks the synapses (synapse set "4" in Fig. 19.2(a)) which are to support the "second enable" input (Sects. 19.1.7–19.1.8).

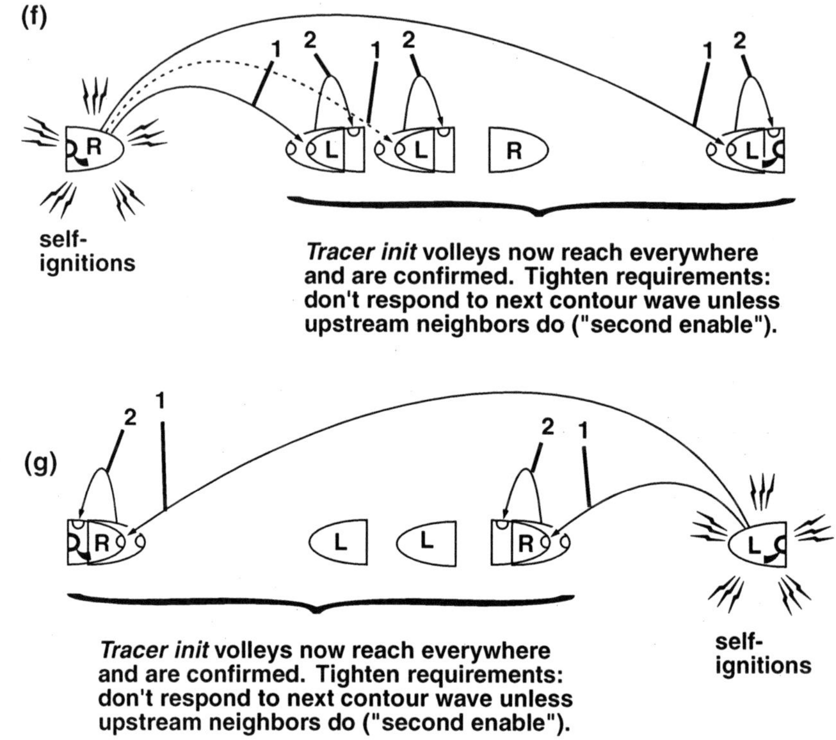

Fig. 19.4(f)–(g) Implied messages during bidirectional contour linkup (Cont.). Tracer waves become continuity detector waves. Up until this point, tracer init volleys emitted from the node at the left set of tracer reports for every contour wave in progress to the right of the node. As mentioned in Sect. 19.1.7, this undermines the continuity detection, since the base node is likely to receive a continuous shower of tracer reports until (and beyond) the next (spontaneous) ignition of any node at the right, even if the node is on another contour. This is because the several tracer waves which are in progress at any moment will, together, tend to mask the gaps in the tracer shower. In order to make sure that there is only one contour wave in progress at any time, a "second enable" requirement is now imposed

As illustrated in Fig. 19.6, the cells at this point do not know whether the incoming volleys are from the same contour (as in the present drawings) or from different contours; they mark the synapses from all repeating volleys, right ones and wrong ones alike, and treat them all as *tracer init* volleys.

The *second enable* signal is simply a short-axon copy of the tracer report spike from the nearest neighboring contour cells which were supposed to fire right before the cell (see synapses marked in (d)–(e)). The only cells exempt from the second enable requirement are the ones within short-axon range of the base node, which are fully enabled by the *tracer init*; as a result, every "second enable wave" begins at a node. A contour wave already away from the node when receiving its *tracer init* would, in order to produce tracer reports, have to be preceded by a *second enable wave* also arising in mid-interval, which is not possible.

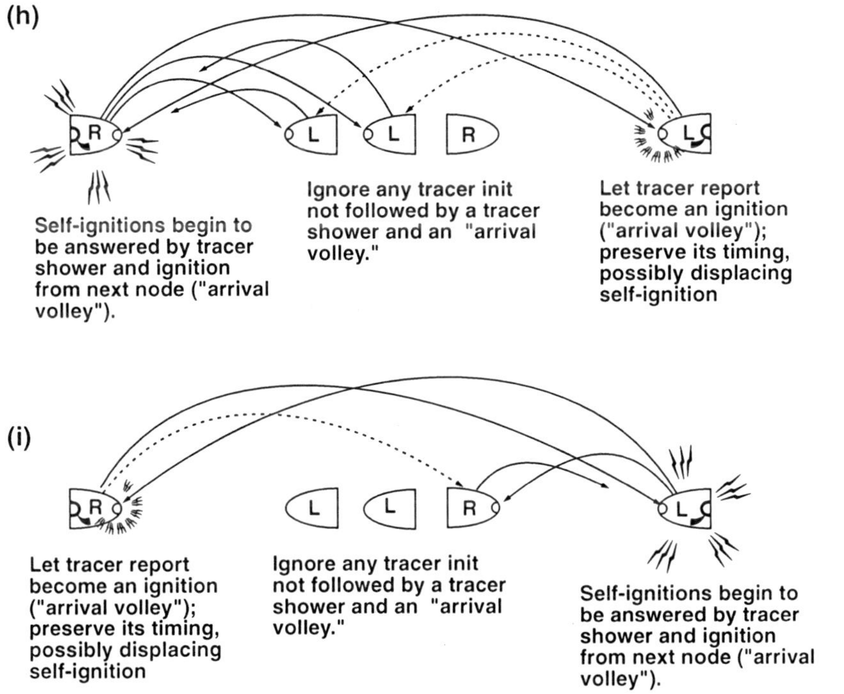

Fig. 19.4(h)–(i) Implied messages during bidirectional contour linkup (Cont.). The "arrival volley" must be timed well to the tracer shower. The final step of the tracer action is the volley issued from the next node when the traced contour wave (and the tracer wave) reaches it. The volley is recognizable to the base node, because it matches a set of (spontaneous) volleys whose synapses are already marked (Fig. 19.3(b)). The continuity information resides in the relative timing of the tracer init, the tracer wave (which must not begin before the arrival of the underlying contour wave), and the arrival volley (which must consistently follow the same cells firing in the tracer wave)

Like the chain of cells between nodes, the cells of the next node are subject to the *second enable* requirement, with the result that, after a tracer wave reaches the node, it causes immediate response which quickly spreads to produce an ignition (*arrival volley*) audible to the base node.

The arrival volley is a reliable indication of continuity, all the way from node to node, provided that it follows upon the tracer shower (as in Fig. 19.3(b)), and that it (frequently enough) follows the shower precisely, and without being held up by the refractoriness after its last self-ignition. Accordingly, the self-ignition "timer" must (ideally) be stopped when the tracer init arrives, until arrival of the tracer wave, which gives the assembly more time to recover.

Once the tracer waves are reliable, and have gone through enough repetitions, they in conjunction with the arrival volley can be used by all cells in the chain for filtering out the ignitions coming from other contours (Fig. 19.6). Other contours run contour waves to link up their nodes, same as the contour under consideration

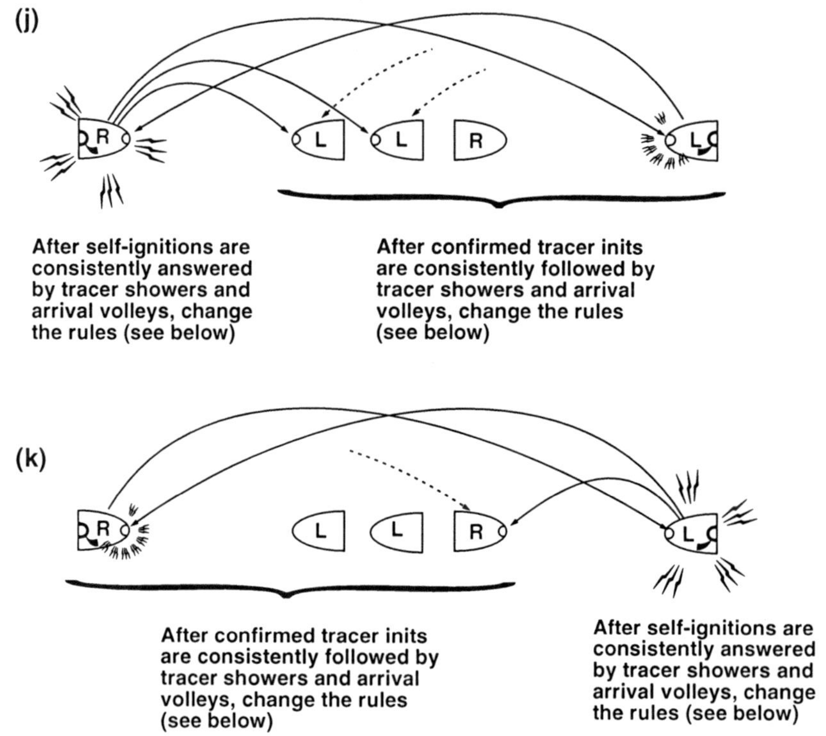

Fig. 19.4(j)–(k) Implied messages during bidirectional contour linkup (Cont.). After continuity is established, tracers are no longer needed. After some repetitions of the sequence made up of tracer init volley, tracer shower, and arrival volley, the validity of the sequence becomes confirmed to all cells of interest, namely the cells in the nodes and the cells in-between nodes. The cells in-between are of no further use. Their role has been to support the communication between the nodes, and tell them that they were on the same contour, and that role is now fulfilled. The nodes both know that they are on the same contour; they have created a group of their own cells reaching the other node and reached by it; and have marked out the synapses from the other node on their cells. The next step is to make a mode transition, to create an active link between the nodes

here, but since each node produces tracer init volleys on a random schedule under its own control, its cells have a basis for filtering out the crosstalk.

The neurons in the nodes formerly receiving input from the tracer waves disable the synapses formerly marked out for them; so it can be said that the nodes thereafter ignore all inputs except for the ones coming from nodes. In addition, they change their responses to the volleys from their opposing nodes to an immediate passive response, causing a local ignition as nearly simultaneous with the incoming volley as possible, and without regard for the underlying contour waves. The nodes still pay attention to the contour waves, but their role is now reduced to one of node maintenance (Chapters 16, 17, 18).

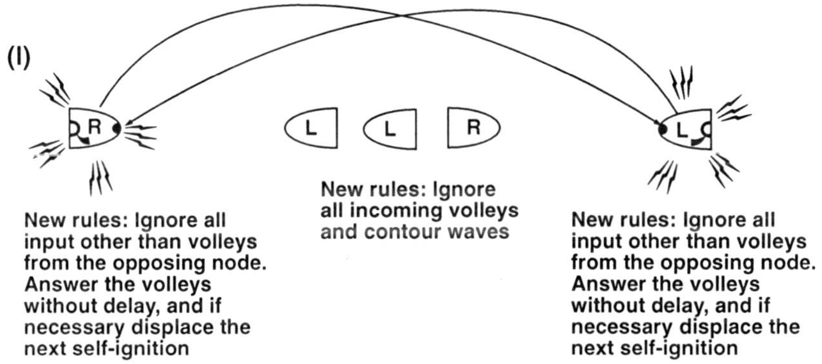

Fig. 19.4(l) Implied messages during bidirectional contour linkup (Cont.). Thereafter, only the nodes fire, emitting self-ignitions or responses. The detection of continuity is the signal for mode transition which ushers in a new set of rules. Under the new rules the neurons with synapses marked out on them become bridgeheads, and the bridgeheads on the two nodes become a single cooperating unit. The neurons between nodes do not record ignitions coming from the nodes anymore, or respond to the underlying contour waves

The end result is that both bridgeheads run self-ignitions independently of one another, while also passively responding to the self-ignition volleys they receive, which is the functional mode earlier described as *active link* operation. The implication is that the active link is engaged in continual and mutual echolocation between the nodes.

The neurons located between the nodes detect the new situation by the fact that volleys from opposite directions now arrive to them in nearly simultaneous pairs. In response to the change these neurons stop running tracer waves, and disable the synapses formerly marked out for the *tracer init volleys*, the *second enable inputs*, and the *arrival volleys*, in other words all equipment formerly devoted to the running of tracer waves.

19.10 Suppressing Tracer Waves Beyond the First Node They Encounter

The intent of the tracer run and subsequent linkup is to create links between adjacent nodes.

The reach of the *tracer init* volley is not limited by the first node, and there is (probably) no clean-cut way to cancel the *second enable* machinery at the next node; therefore the issue arises that the tracer waves will run on past the node they are intended for locating, and cause linkup with nodes beyond it as well. The overrun issue can be addressed by the continuity detection protocol at the base cells, as described in Fig. 19.5.

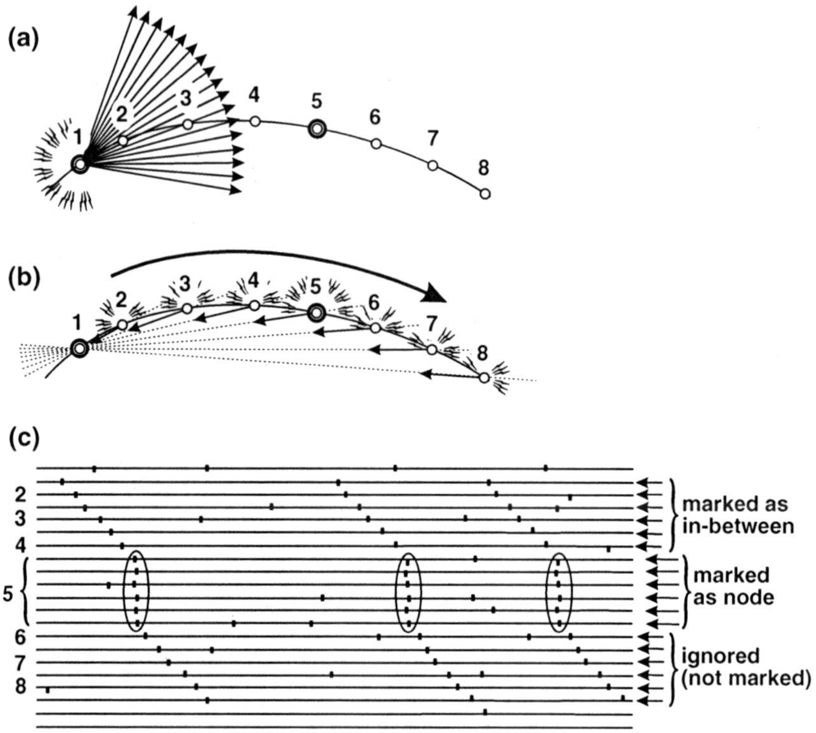

Fig. 19.5 Recognition of tracer waves returned from beyond a node. Nodes are at locations 1 and 5, the other locations stand for neurons responding individually. A tracer init volley, shown in (**a**), causes responses from all cells up to a certain distance to the right of the base node, as shown in (**b**), including some located to the right of the next node. The responses, as received at the base node (location 1), are shown in (**c**). As seen, the tracer reports returned from locations 6, 7, 8 can be distinguished from the rest by the fact that they arrive after the volley from the node at 5, which arrives to synapses already marked. On the basis of the timing, the synapses to which the late-coming spikes arrive can be un-marked

19.11 Recognition of Crosstalk Between Two Contours

Contour cells send their outputs into wide angular regions, and crosstalk between different contour strings will tend to confound the continuity detection, as is shown in the example of Fig. 19.6(a)–(c). It is not quite obvious at first glance that base cells can even distinguish a tracer shower originating from their own color patch from tracer showers which originate from other color patches.

In fact the following operation would not be possible if self-ignitions occurred periodically; it is only made possible by the random scheduling of tracer initiation volleys.

A series similar to Fig. 19.6 can also be drawn for two elongated color patches placed end-to-end, where the nature of interference is slightly different, although

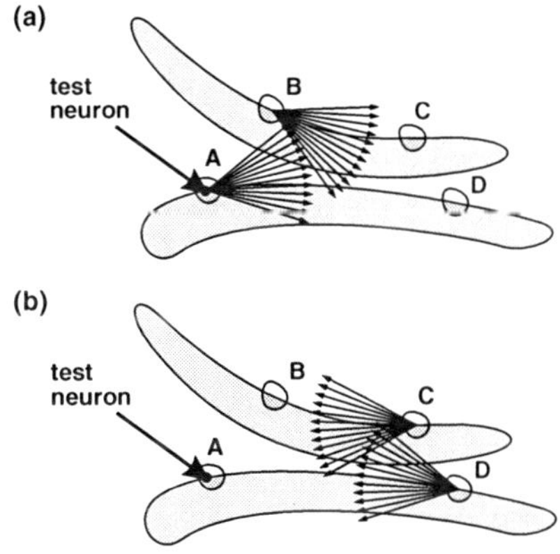

Fig. 19.6(a)–(b) Recognition of volleys coming from another contour. Two color patches situated to be vulnerable to crosstalk. Two patches are shown with the same color scheme, such that the tracer init volleys are not prevented from acting upon neurons of the wrong contour, and base nodes are not prevented from receiving tracer showers initiated (and properly running) on the wrong contour. The question is how the nodes of one color patch avoid linking up, by mistake, to nodes of the other color patch. The answer is, in essence, that the second enable processing limits tracer waves to one per tracer init, running always on the right contour; and the random schedules of the tracer init volleys, different on different contours, enable the base node to separate the right tracer showers from the wrong ones

the principle remains the same. There, too, the discontinuity in the tracer shower becomes established when subjected to the test of confirmation loops.

Echoes from the wrong contour will be out of sync with the interrogating volleys. The illustration in (c) attempts to spell out the same reasoning in more detail.

Nodes A and B send out *tracer init* volleys independently of one another, on two different random schedules. Since the volleys go out on fairly wide angles (arrows fanning out), both *tracer inits* reach some neurons from the wrong color patch. Similarly the *arrival volley* responses from the nodes C and D, emitted when the contour waves arrive to them, reach both A and B.

In the examples shown, the close relation between tracer runs and arrival volleys is of no help, because all three tracer runs (T1, T2, T3) are continuous with an arrival volley (AV1, AV2, AV3), including the two coming from the wrong contour. The difference between right and wrong is that only the *tracer run* from the right contour, T2, follows upon a *tracer init* volley in which the test neuron participates, and is triggered by the first contour wave after the volley, which is also local to the test neuron.

That one difference is sufficient, though, when taken together with the fact that in randomly repeated occurrences the spatial continuity is present each time. Based on it, it is possible to mark the synapses bringing the tracer waves and the arrival volleys from the right contour (T2 and AV2) and leave the others unmarked (arrows at right).

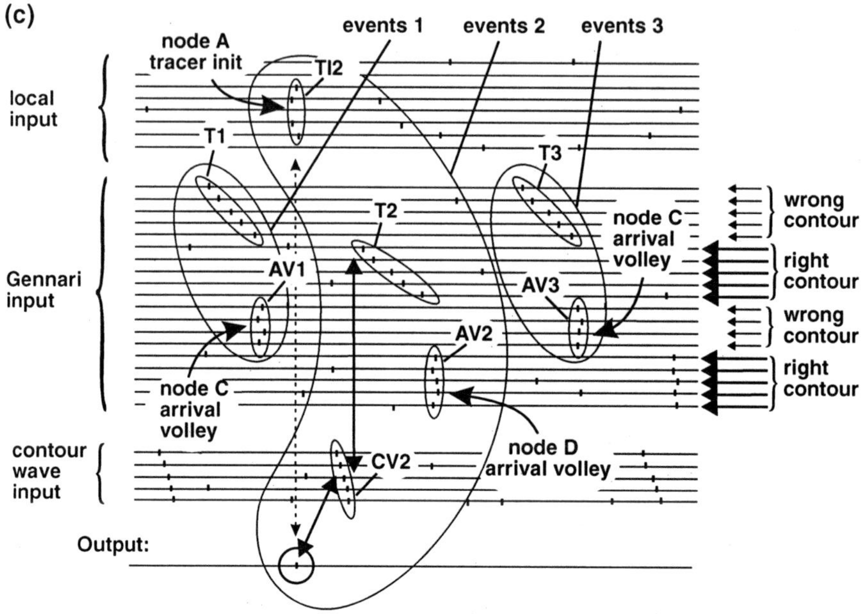

Fig. 19.6(c) Recognition of volleys coming from another contour (Cont.). Appearance of tracer responses from the two contours. The spike train drawings show the inputs and output of a test neuron located in node A. The crucial point to remember is that the Gennari fibers from the "right contour" and the "wrong contour" do not belong to physiologically distinct neuron pools; the only thing making them different is the momentary retinal location of visual images. The test neuron is faced with the task of detecting the difference and marking the synapse sets accordingly. The drawing shows three sets of events, 1, 2, 3 (each circled), of which events 2 are from the right contour, and events 1 & 3 are from the wrong one

As a separate issue, one may note that in looking at real-world objects it often happens that one object is in front of another, partly blocking it from view and showing it only as two disjoint pieces. Although the visual system can usually recognize the latter as being a single object, the solution is believed to lie outside the primary visual cortex, and is beyond the scope of this model.

Chapter 20
Using Existing Links to Make New Links on the Same Contour

20.1 Outline of Using Two Links to Make a Third Link on the Same Contour

20.1.1 Relation "A on Same Contour as B" is Transitive; But There is a Catch

As was mentioned, before nodes can convey the shape of a contour through directional co-ignitions, they must first form directional links with each other, and before linking up to any node they must first be sure it is on the same contour as they themselves.

After every node knows that the nodes to its right and left are on the same contour, the nodes are no longer totally ignorant of the rest of the contour, but the "ignorance problem" is still only partly solved. The nodes still do not know that the nodes beyond their neighbors are also on the same contour.

In general, we, as people looking at the overall situation, can make the deduction that when nodes A and B are on the same contour, and nodes B and C are on the same contour, then nodes A and C are also going to be on the same contour. In other words, we know that the "same contour" relation is *transitive*.

However, in a brain context, the linkup decision has to be made *locally* at each node about to form a link, and that node must first know that the other node involved in the linkup is on the same contour.

And here is the catch. Transitivity of the same-contour relation does not help in the local decision whether to make a link, because in this case it relies on combining two antecedent statements known at two different locations. Location A may know that location B is on its contour, and location B that location C is on its contour, but when A needs to make a linkup decision the only thing that matters is what location A knows.

When the objective is to form bi-directional linkage, as it is in all the cases of interest here, the same considerations apply to both nodes participating in the linkup; they each must first know that the other is on the same contour.

It is this requirement to transfer knowledge between nodes which leads to the need for the procedural somersaults described in Figs. 20.1(a)–(h), 20.2(a)–(n), all of them depending on one node "understanding" what another node means by its

C.R. Legéndy, *Circuits in the Brain*, DOI 10.1007/978-0-387-88849-1_20,

ignitions. The secret is invariably that the receiving node must recognize the overall (global) situation, since that is what holds the key to the local interpretation of the volleys.

20.1.2 Linkup of Two Nodes Must Start from a Third, with Links to Both

When three nodes on a contour are connected by two links, and it is desired to close the triangle, the process must by necessity begin from the node which has links to both of the other nodes, because only that node knows that it is on the same contour as both of the others. In other words when the two existing links are a link between nodes A & B and one between nodes B & C, the linkup between A & C must be initiated from node B. For reference, in this configuration, I will call B the *fulcrum* node of the linkup, and call A and C the *satellite nodes*.

The first thing to note is that just because the *node* knows of both existing links, it does not follow that any neuron in the node does. It is more likely that the node has some neurons (in one bridgehead) which know about A and some others (in another bridgehead) which know about C.

Accordingly it is necessary, in addition to the "node-level" description (Figs. 20.1, 21.1), to include a neuron-level description (Fig. 20.2), where it is made sure that at some point the required knowledge is concentrated in some neurons which, at least temporarily (while the necessary decisions are made), know about both.

The problem of finding such neurons is not hard to solve, as will be seen in a moment. The more tricky question is a different one: how can node B know that A does not yet have a link to C?

The latter question, which goes beyond the usual two-element relations encountered in this model, is addressed in connection with the general question of completing a triangle of links (Figs. 21.1 and 21.2).

For the sequences of the present chapter we bypass the question by concentrating on a series of linkup steps where prior to every step (except possibly in the last) it is known that there is no link between the satellite nodes. Such a series is obtained by envisioning (for instance) a curved contour with two corner nodes at the two ends and a series of other nodes in-between, and single-stepping through the sequence of nodes from one corner node to the next (Figs. 20.1 and 20.2).

20.1.3 Three-Node Ignitions

Let us consider the situation where links exist between A&B and between B&C, and where node B is chosen to be the fulcrum of a linkup between A and C. The next question is how do the neurons in B proceed?

The solution used here utilizes the fact that both the A–B link and the B–C link are *active* links, in the sense of Sect. 16.2, meaning that they repeatedly perform co-ignitions on independent random schedules. The trick is to fuse the two links into a "three-node active link," in which the A–B link passively follows every co-ignition of the B–C link and vice versa. This is accomplished by creating a two-way linkage between the two bridgeheads in node B.

The resulting three-node co-ignitions can be used to *tell* nodes A and C that they are on the same contour, provided the cells in A and C *know* (though genetics and chemistry), as we assume they do, that *whenever groups of direction-coded cells or contour cells emit simultaneous volleys, from two mutually remote sources and on a random schedule, then the groups are on the same contour.*

20.1.4 Making the Triple Ignitions Reach the Satellite Nodes

In this way the triple ignitions can in principle tell the cells in A and C that they are on the same contour, but in practice there is a problem that involves reaching.

The triple ignitions at this point are confined to four bridgeheads, one in A, two in B, and one in C, all of which are small sets of cells hand-picked to reach the node they are aimed at; they don't necessarily reach any other node. In particular, there is no reason to assume that the C-component of the triple ignitions will reach A, or that the A-component will reach C.

In order for the B-initiated ignitions of C to reach A, they have to be made "omnidirectional," by being made to spread to all the long-axon neurons in C which could potentially reach A. Since the cells in C do not initially know where A is, this means that, until contact is made, the simplest solution is one that causes all cell groups in C (and similarly all cell groups in A, for reaching in the opposite direction) to follow, passively, the triple ignitions which show themselves to them as the ignitions of the local bridgehead co-igniting with B. In other words reaching between A and C is assured if A and C emit *omnidirectional* volleys as their contribution to the triple ignitions.

20.1.5 The Issue of Limiting the Search Volleys to a Range of Directions

The omnidirectional volley is the simplest kind of "search volley" (meaning the tentatively widened volley intended to reach from A to C or vice versa), but it is likely that the brain uses a more efficient form of search volley, because activating every direction-coded group in a node puts more firing into the cortex than is probably necessary.

The more reasonable strategy is first to send search volleys into a narrow range around an initial guess, and then gradually widen the range until the desired node is reached. (The obvious choice of initial guess is the direction pointing from the satellite node to the fulcrum node.)

The problem is that a strategy of using "gradual widening of search volleys" amounts to finding direction-coded cells with effective fields which are "only slightly different" from those of a given group of cells, thereby requiring a built-in distance function ("metric") attached to pairs of direction-coded cells.

We recall that a similar requirement for a metric on synapse sets or neuron sets was already encountered above, in connection with the warm-up of simple cells (Fig. 12.2) and the tracking of visual images suffering distortion (Sect. 16.5). As in the previous cases, I assume here that the brain finds a way to solve the problem of gradually widening the search volleys around an initial guess, and avoid the need for the wasteful and noisy solution of sending the volleys to all directions and all distances.

However, for simplicity, I will stay with the wasteful but easy solution of the "omnidirectional volleys" in the descriptions that follow.

20.1.6 How Do Nodes A and C Know that They Are Supposed to Link Up?

Turning to the generation of the omnidirectional volleys, it can said that, while the satellite nodes A and C are believed to be capable of issuing such volleys when necessary, they will not do so, unless they know that they need to.

This takes us to the next question: how can the fulcrum node B tell the satellite nodes A and C that they are to try and link up to each other?

Here, the solution we choose makes use of the fact when the fulcrum node ties its two bridgeheads together to create triple ignitions, these ignitions will occur at a noticeably higher rate than the former co-ignitions. The reason is that all the former B–C co-ignitions will spread to A, and all former A–B co-ignitions will spread to C; so the rate of triple ignitions will be roughly the sum of the two former rates, or very roughly twice the former rates.

The increased rate of co-ignitions is detectable at the satellite nodes A and C, and can serve to tell them to take the first step in the linkup by initiating the omnidirectional volleys.

20.1.7 The Beginning of the Bridgeheads of an A–C Link

Accordingly, when node A detects the increased rate and begins sending out omnidirectional volleys, it can be assumed that sooner or later C will also start sending them out and A will sooner or later be in a position to detect the volleys coming from C. Similarly, C will soon be expected to detect volleys from A.

In order to utilize this fact for linkup, all cells participating in the omnidirectional volleys must also seek to *detect* volleys coming from any *new* (long-axon) source which is in sync with the triple ignitions, and whose volleys appeared around the time the omnidirectional volleys were started.

The neurons which detect such new volleys will know (Sect. 20.1.3) that the source of the new volleys lies on the same contour as they are themselves. Because they know they are participating in a linkup, they will also know that the source of the new volleys is the node with which they are supposed to link up.

As was seen in Chapters 16 and 17, the bridgeheads of links must be ignitable groups. This means that the cells of the two nascent bridgeheads, up until now identified only by being reached from the opposite nodes, must next form ignitable groups.

20.1.8 The Challenge of Making the Bridgeheads Ignitable

But there is a problem. Making the candidate bridgeheads into an ignitable group amounts to creating ignition support links between cells identified by nothing else than that they are both reached by volleys coming from a remote node. And passive bombardment by the volleys alone is not enough for creating the links because the cells only know that they themselves are members of the reached group; the bombardment does not tell them which of their neighboring neurons are similarly reached.

The creation of ignition support links for the bridgehead requires a way to tell the neurons which of their synapses comes from other members of the desired group. The difficulty is caused by the fact that in the omnidirectional volleys all direction-coded cells in the node fire at the same time, and their firing is in sync with the firing from the remote node which singles out the selected cells (see ignition sequences in Fig. 20.2(h)). Therefore the linked timing between presynaptic and postsynaptic cells will be automatic, and the fact of simultaneous firing alone will not convey any information as to causation.

The solution we choose uses the assumption that the direction-coded cells convey information through the *labeled lines* principle, which means that by the same method which the V2 cortex uses to deduce directional angle from their volleys, their neighboring V1 cells, the ones which are reached from the opposite node, can also use for recognizing when their directional angle is the same as their own.

Although shared directional angle is no assurance that the other cell is reached by the opposing node, it is a good enough indication, because of the assumed reciprocity of contacts, to justify the tentative formation of ignition support contacts.

The synapses reinforced on the reached cells in this way will form a superset of the synapses which will eventually survive, because angle of relative direction is shared between more cells than actually reach a neighborhood (Fig. 8.1), but the next steps will serve to eliminate the extra contacts.

20.1.9 Gradual Growth of the New Bridgeheads

It is important to note that initially the set of cells reached from the remote node is not necessarily expected to be large enough to form an ignitable group, because the nodes which are meant to be interconnected may be quite far apart (some millimeters, and in primates even some centimeters apart). But as the nodes grow, the set of cells reached is expected to gather more members (Fig. 17.1).

Now, cell groups need to be of a certain minimum size in order to be ignitable, (Legéndy, 1967), and accordingly the groups of reached cells may initially be too small to ignite (or in fact non-existent). By the same token, since the nodes grow, and more and more of their cells are reached from the other node, at some point the groups are expected to grow to a size where they are ignitable.

The mutual contacts created between the reached cells will support ignition when the group is large enough. Since the objective is to turn the cell groups into bridgeheads of an active link (see 16.2), as soon as the groups become large enough to ignite, they will start performing self-ignitions. This means that, in addition to following the triple ignitions, the new groups will repeatedly go into instability and send out volleys at times determined by themselves, out of sync with the triple ignitions (Fig. 20.2(i)).

The new out-of-sync volleys emitted by one satellite node will in time be detected by the other.

20.1.10 The Cessation of Omnidirectional Volleys

The omnidirectional search volleys are needed during linkup, but they send a great deal of firing into the surrounding cortex, which interferes with the processing on other contours. Therefore it is important that they are not maintained any longer than necessary.

Stopping the omnidirectional volleys amounts to stopping to follow the triple ignitions. The cue for doing so is different for different cells.

It is easy for the cells in A and C which are reached from the opposite satellite node, and are now executing self-ignitions. They have two sources of information at their disposal. First they know that they themselves are numerous enough to support self-ignitions, which means that their group is reached by enough fibers from the opposite node. Second, they are able to detect whether there are out-of-sync volleys reaching them from the remote node. If there are, that means that the remote node contains enough cells reached by their own ignitions to form an ignitable group. Accordingly, when both sources of information tell them that the linkup is viable, they know that the omnidirectional volleys have served their purpose, and stop following the triple ignitions.

The situation is different for the cells not reached from the opposite satellite node. The fact of failure to be reached is, in itself, not a good cue for disconnecting, since these cells do not know whether they would be reached as soon as the nodes have grown large enough or would never be reached because their directional coding is off, and they are missed by the fibers from the other node altogether.

To these cells, the first cue is when they detect the local self-ignitions coming from the newly ignitable cells near them. The self-ignitions are recognizable by being out of sync with the triple ignitions, and they carry an indication that the number of cells reached from the remote node is sufficient to support the self-ignitions. But from the independent local self-ignitions alone they do not know whether the local node also reaches the remote node. If it does not, the omnidirectional volleys should be continued for a while to let the growth of nodes achieve better reaching.

Accordingly, the second and final cue for them to disconnect from the triple ignitions is when they detect that the out-of-sync local cells no longer follow the triple ignitions. When they no longer do, it means that they have detected the self-ignitions coming from the remote node, which in turn means that the reaching is bidirectional. They too then stop following the triple ignitions and, lacking any other signal to follow, fall silent.

Once both groups of local cells have disconnected themselves from the triple ignitions, the node no longer emits omnidirectional volleys (Fig. 20.2(j).

20.1.11 Setting Up Mutual Excitation Between Nodes A and C

At this point there are not yet any strong contacts between nodes A and C; the synapses are marked only, and their role is merely information transfer.

The next step is for the two nodes to start following each other's ignitions, thereby forming an "active link." This will enable the two bridgeheads to interact in the way described in Chapters 16 and 17. The nodes will then be able to track each other as the contour drifts, use "echo detection" to monitor their bidirectional communication, and shed any cells not useful to the linkage.

20.1.12 The Next Step Is to Separate the A–B Link and B–C Link Again

After the bridgeheads on nodes A and C have been established, and the linkage between A and C is active, it is time to disassemble the rest of the special arrangements made in support of the linkup. The first step, the stopping of the omnidirectional volleys, was already taken.

The next step is the stopping of triple ignitions altogether, which means dissolving the linkage between the bridgeheads inside B (the B bridgehead of the A–B link and the B bridgehead of the B–C link), and returning the two bridgeheads to their original independent operation.

20.1.13 How Do the Bridgeheads in B Know to Undo Their Linkage?

The question is how the bridgeheads inside B find out that the link between A and C is now built and operational?

It will be remembered that the bridgeheads which must know this, and consequently that they should drop their linkage, both reside in node B, but the reason they should drop the linkage (completion of the A–C linkup) is made up of changes all residing outside node B.

The answer is that, although the bridgeheads of interest do not know of the successful linkup, they are bidirectionally linked to cell groups which do. In other words the B bridgehead of the A–B link does not know of the change but the A bridgehead of the same link does, and similarly for the A bridgehead and C bridgehead of the B–C link. The bridgeheads that know of the change know it by the fact that they are no longer being followed by other cells in their node.

However the fact that the A bridgehead of the A–B link knows about not being followed does not mean that the B bridgehead also knows, and it is the B bridgehead which needs to know it (and similarly for the B–C link).

So the next question is how the A bridgehead of the A–B link can tell the B bridgehead that it should disconnect from the other B bridgehead? As before, the problem is the "axonal bottleneck" in neuronal communication, in other words the fact that the neuron knows more than it can say, since all it can do in way of saying things is to fire or not to fire.

The solution chosen here makes use of the fact that a bridgehead of a link can always tell whether the other bridgehead follows its ignitions. The ignitions from the other bridgehead can reach it, and therefore its neurons can detect the presence or absence of synchrony with their own ignitions (Fig. 16.1).

Therefore the A bridgehead of the A–B link can communicate the fact of its changed situation by ceasing to follow the ignitions from B. The B bridgehead in turn is in a position to know that under the circumstances the cessation of following means that node A has successfully formed a bridgehead toward node C, and that it is time for the two B bridgeheads to drop their mutual dependence. (The same holds for the bridgeheads of the B–C link.)

In other words, the two nodes, A and C, independently tell their bridgeheads in B when they can drop their connection with the other bridgehead in B. When both bridgeheads receive and interpret the message, the triple ignitions are stopped; the A–B link and the B–C link resume their independent action.

The assumption that linked nodes communicate to others by stopping to follow them, like the assumption of communicating through increased rate of ignitions, was chosen here merely because I could not think of any better solution that utilized neuronal firing and nothing else, but Nature may well have found a better solution.

20.1.14 Restoring the A–B Link and B–C Link

It still remains to restore the A–B link and the B–C link, disrupted in order to tell node B to disengage its two bridgeheads.

One more question remains: how is the A bridgehead of the A–B link and the C bridgehead of the B–C link to be told that the two B bridgeheads have successfully disengaged themselves from each other, and that therefore the satellite bridgeheads can resume passively following the B bridgeheads?

Here the solution chosen is similar to one of the previous ones and makes use of the changes in co-ignition rates. When the two B bridgeheads are mutually disengaged, their co-ignition rate is reduced by roughly half, putting the A and C bridgeheads in a position to know that the disengagement is complete and their operation can be restored.

20.1.15 Why Not Just Start a Free-for-All of Echolocation?

It may be suggested that instead of all the complex steps listed above, a simpler way exists for the nodes to link up, by some variant of old-fashioned echolocation.

The first problem here is that before echolocation can be used in linkup, both partners of any potential echo dialog must agree on a signal telling them to go into echolocation mode. Echolocation in the radar or sonar context does not require cooperation of the target, since the laws of physics will make sure that the target issues an echo. By contrast, in the neuronal context, responsiveness to an input must first be arranged.

However, even if this is achieved somehow, it is clear that if all nodes on all contours looked for linkages by independently emitting omnidirectional volleys, they would fill the cortex with firing and drown out all signals in noise.

The triple ignitions greatly reduce noise, since all triplets are guaranteed to be on the same contour, and the self-ignitions driving the different triplets, usually on different contours, are out of sync with each other. Their disadvantage is that they can take quite long to link up a whole contour's worth of nodes, because to link up two corner nodes separated by five pass-through nodes requires five three-node games, one after another. Accordingly another possibility has also been outlined (Fig. 22.1) for use on contours having relatively few nodes.

The alternative process, like the one just outlined, begins after all nodes have linked up to their neighbors by means of contour links. The difference is that in the all-to-all linkup, triple ignitions spread to quadruple ignitions and larger ones, each time initiating omnidirectional volleys and linkup to all co-igniting nodes. The spreading process continues until it either succeeds in linking up the whole closed contour, or overwhelms the participating cells with noise, is rolled back, and initiates a sequential triple ignition protocol (Figs. 20.1 and 20.2).

20.2 Extending a Long Link to the Next Node on a Contour

The next sequences presuppose that all pairs of adjacent nodes are already tied together by contour links, and describe the way in which the system can use these links, and their guarantee that they tie together nodes on the same contour, to create further links, which are then also guaranteed to tie together nodes on one contour.

In the sequences of this section we "single-step" through the string of pass-through nodes between two corner nodes, and thereby gradually extend the linkage in a succession of three-node plays.

In each step it is assumed to be given that a corner node already has a directional link to a node several steps removed from it, and therefore already knows that it is on the same contour, and the objective is to use this link, together with a contour link, to extend the same-contour knowledge to the next node.

If we were to follow the schema of mathematical induction properly, we would also need to spell out the description for first step in the trivial case where the first contour link of the interval is supplemented by a directional link connecting the same nodes, but I will not bother with that, to save space. In addition, since the subject here is a finite sequence, and the last step also differs from the rest (it creates a link between two corner nodes), it should also be spelled out in a more complete presentation, but I will skip this step, too.

In the sequences shown in Figs. 20.1 and 20.2, the corner node to which linkage is made, referred to as the "anchor node," of the linkup, is numbered node 1. In each step the anchor node knows, to start with, that it is on the same contour as node n, and the step uses the contour link between nodes n and $n+1$ to tell the anchor node that it is also on the same contour as node $n+1$ (and tell node $n+1$ that it is on the same contour as the anchor node). In order to shorten the notations somewhat, the drawings arbitrarily use 6 and 7 instead of n and $n+1$ for the serial numbers of the two consecutive nodes. As will be noted, nodes 1, 6, 7 take the place of nodes A, B, C in the outline of 20.1 (and the *triangle closing* discussion of Chapter 21).

20.2.1 Node-Level Description of the Linkup Step

It is useful to go through the steps of the linkup in a schematic form, parsing the procedure down to the level of nodes only, as in 20.1, and (more or less) ignoring the need for information transfer inside the nodes. The verbal references to the cues in question could be supplemented everywhere with spike train drawings along the lines of Fig. 16.1 or Fig. 19.3, but those are straightforward and are omitted.

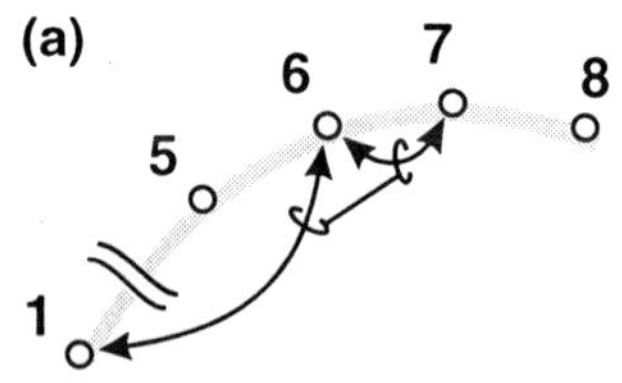

Fig. 20.1(a) Extending a long link, in rough outline. Node 6 initiates new linkup by fusing two old links. Node 6 detects completion of its link to node 1 (at the end of the previous iteration of the process). Then node 6 ties its just-completed direction-coded-cell bridgehead facing toward node 1 to the already existing contour bridgehead facing toward node 7 with a strong excitatory link. The result is that thereafter whenever any one of nodes 1, 6, 7 ignites, the other two will passively follow (triple ignitions)

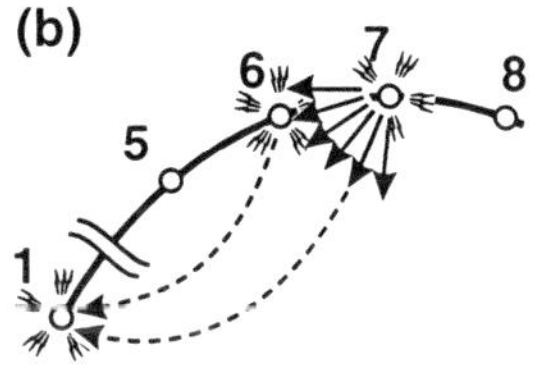

Fig. 20.1(b) Extending a long link, in rough outline (Cont.). Nodes 6 and 7 hammer on node 1 from the right. Because of the triple ignitions, which combine the ignition rates of the former 1–6 and 6–7 ignitions, node 7 receives volleys from node 6 about twice as often as before, and interprets the greater ignition rate as a command to start linkup. As first step of the linkup, node 7 makes its own ignitions less precisely aimed (omnidirectional volleys) with the result that they no longer reach only node 6, but reach node 1 as well (nodes 6 and 7 "hammer" on node 1 from the right). At the same time, node 1, which also detects the greater frequency of volleys coming from 6, stands by to detect volleys from any new source in sync with its own ignitions

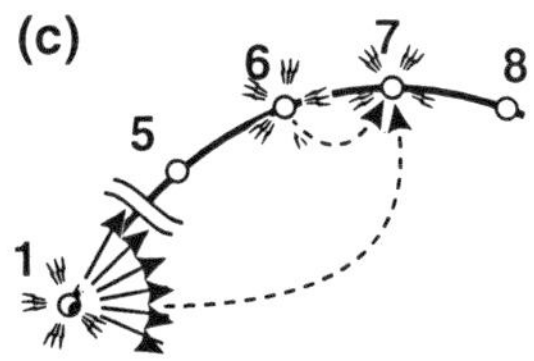

Fig. 20.1(c) Extending a long link, in rough outline (Cont.). Meanwhile nodes 1 and 6 hammer on node 7 from the left. The same steps are made on the other side: Node 1, detecting the greater frequency of ignitions, makes its own volleys omnidirectional, and hammers, together with node 6, on node 7 from the left. In response to the same cue, the greater frequency of volleys, node 7 stands by to detect volleys from any new source in sync with the triple ignitions

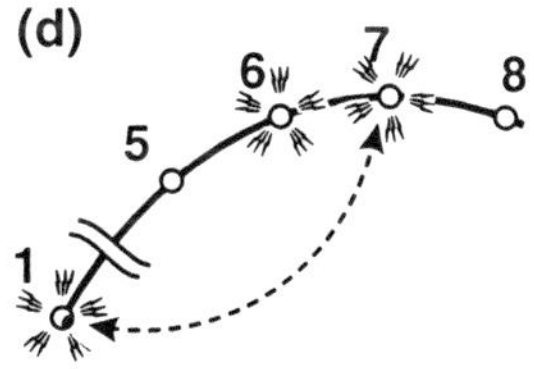

Fig. 20.1(d) Extending a long link, in rough outline (Cont.). Cells reached by new volleys form ignitable groups. Neurons in nodes 1 and 7 detecting that they are reached by the other node's volleys link up with nearby cells having similar direction coding, since those are likely also to be reached from the opposing satellite node; the purpose being, at each side, to turn the cells reached from the opposing node into an ignitable group. This may take a while, since the nodes may have to grow before they can reach each other (Chapter 16). Once reaching is achieved, and the groups become ignitable, both groups start autonomous self-ignitions

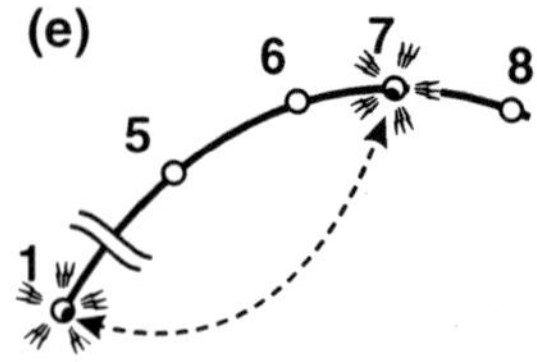

Fig. 20.1(e) Extending a long link, in rough outline (Cont.). Nodes 1 and 7 stop their omnidirectional volleys. The newly combined cells in nodes 1 and 7, after starting their self-ignitions, stop following the triple ignitions, as do the rest of the cells in nodes 1 and 7 (the cells not participating in self-ignitions), with the result that no more omnidirectional volleys are emitted from nodes 1 and 7, only the spontaneous co-ignitions of the two nodes. Next, nodes 1 and 7 develop strong excitatory linkage to one another, thereafter passively following each other's random ignitions, in other words forming an active link (Chapter 16)

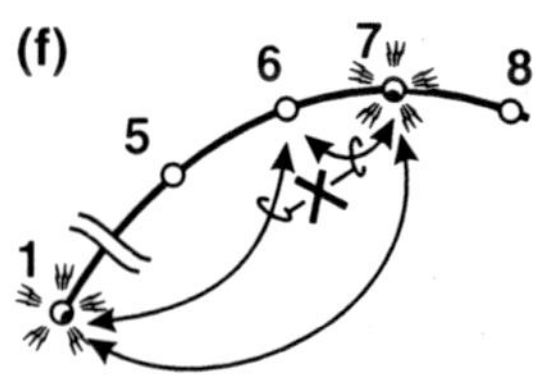

Fig. 20.1(f) Extending a long link, in rough outline (Cont.). Once nodes 1 and 7 are linked up, the triple ignitions are stopped. In node 7, when the bridgehead facing node 6 detects that it is not being followed by the rest of node 7 (in other words that the omnidirectional ignitions have been stopped in node 7), it tells this to node 6, by temporarily stopping to follow its ignitions. The bridgehead toward node 7 in node 6, in turn, detects that it is no longer being followed by node 7, and interprets this as a signal that the 1–7 linkage is complete, the triple ignitions are no longer needed, and it should disconnect itself from the 6–1 bridgehead in node 6. The same steps take place at the node 1 side. The result is that the linkage inside node 6 is dissolved, and the triple ignitions are stopped

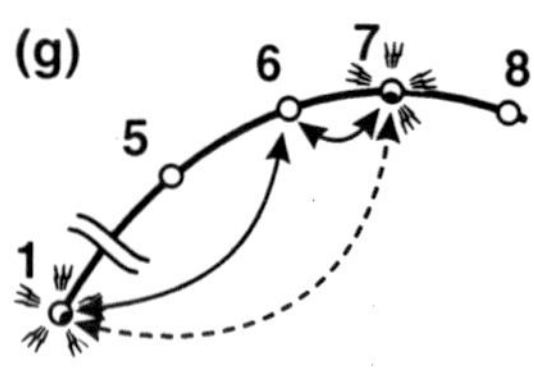

Fig. 20.1(g) Extending a long link, in rough outline (Cont.). Links 1–6 and 7–6 are restored; linkup is complete. Since the ignition rates of the 1–6 link and the 6–7 link are no longer combined, node 7 detects that the rate at which it receives volleys from node 6 has dropped, and interprets this as a command to resume following the ignitions of node 7, re-establishing normal active link operation of the 6–7 link. The same steps take place at the node 1 side. The triangle of links between nodes 1, 6, 7 is now complete, and further corrections of the nodes and links proceed along the lines of node maintenance (Chapters 16, 17, 18)

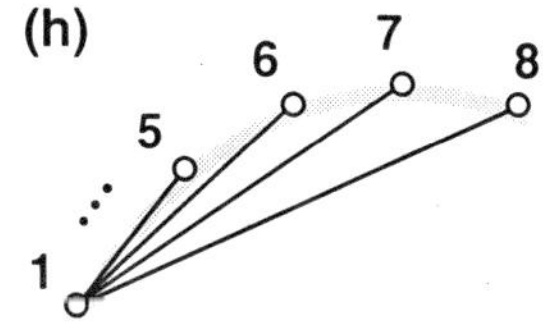

Fig. 20.1(h) Extending a long link, in rough outline (Cont.). End configuration: node 1 linked to all other nodes. After enough repetitions of (a)–(g), the result is a configuration of relative direction links fanning out from node 1 to all nodes on the segment

20.2.2 Description of the Linkup Step in More Detail

The next sequence describes the process sketched in Sect. 20.1 and Fig. 20.1, but in more detail. An attempt is made to sketch the interaction between different nodes, while also describing the interplay between neuron groups within the nodes, down to the individual synapse sets on the neurons. It is not easy to include so many levels of detail and still keep the illustrations self-explanatory; the result may be a little clumsy. I beg the reader's indulgence.

In this sequence only the contour cells and direction-coded cells are included, the kernel cells are left out.

As in Fig. 20.1, node 1 is taken to be a corner point, and is used as the "anchor node" to which relative direction links are being created from all nodes. At the beginning of the run, previous similar runs have already created relative direction links from the anchor node to all nodes up to node 6. The run itself uses the existing link between nodes 1 and 6, and the existing contour link between nodes 6 and 7, obtained through the tracer runs of Chapter 19, both carrying an assurance that they connect nodes lying on the same contour, the two together are used to create a relative direction link between nodes 1 and 7, which consequently also carries the assurance that its nodes are on the same contour.

The straightforward step of replacing the contour links obtained in Chapter 19 by relative direction links connecting the same nodes is omitted, and the steps are shown as relying on the contour links.

For better illustration of the changes occurring in various steps, most ignition sequences are drawn to repeat the same "random" time sequence. Propagation delays are neglected in the ignition sequence drawings; more realistic drawings of co-ignitions would indicate the time lags between ignitions, with the self-ignition always leading (as in Fig. 19.3(d)).

The fusion of two bridgeheads on a node is in effect an "initiate linkup" command by which one node can tell two other nodes, at the other ends of the two bridgeheads, to link up. The other nodes detect the presence of a message by the increase of co-ignition rate caused by the fusion of bridgeheads and know the contents of the message (Sect. 20.1.6). In practice the rate of co-ignitions probably returns to its original value after a while, because the higher rate has disadvantages, but it is maintained long enough to transmit the required information reliably.

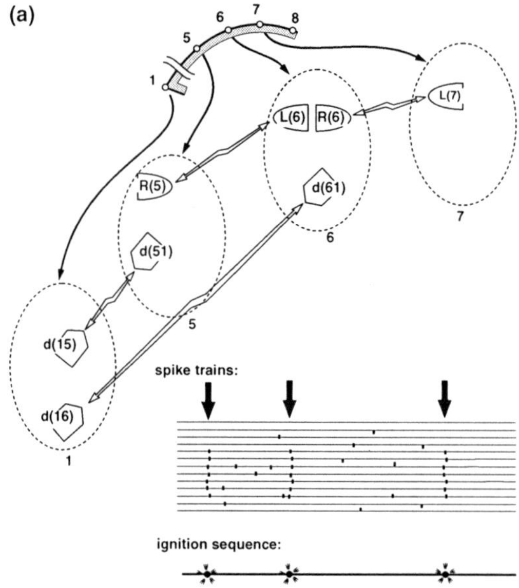

Fig. 20.2(a) Extending a long link while remaining on the contour. Notation. At the *top* of the drawing a contour is shown schematically, with a few nodes drawn in, including the anchor node, numbered 1. A schematically magnified version of the nodes of interest is added below the contour, *circled* in elliptical broken lines. Direction-coded cell groups are shown as obelisk-shaped designs, and the pairs of such groups with mutually reaching axons are shown to be in line with one another for illustration; contour groups are as in previous figures. Direction-coded cell groups are designated with the letter d, with node numbers in parentheses, for instance d(16) is the group of direction-coded cells with cell bodies in node 1, and long (Gennari) axons reaching node 6. Double arrows show active links. At the lower right, the *ignition sequence notation* is recalled (see also Fig. 9.11); it is useful when the purpose is to indicate no more than the timing of ignitions. Slanted arrows on some ignitions on these diagrams (see below) indicate self-ignitions driving the sets of simultaneous ignitions

Nodes 1 and 7 do not know each other's location to start with, since their locations are determined by the retinal image which is controlled by the outside world, and in order to link up, they have to emit "omnidirectional" volleys (see (d)), or at least volleys going out in all directions that could potentially include the other node (Sect. 20.1.5).

In node 1 this means marking and strengthening all synapses coming from d(16), the local group which increased the rate of its volleys, to the direction-coded groups to be activated (denoted d(1x)); similarly in node 7, all synapses from L(7) to the corresponding direction-coded groups (denoted d(7x)).

The direction-coded cells, when going into their "initiate linkup" mode, must not only create omnidirectional volleys, but must also become sensitive to the arrival of "new" volleys in sync with the triple ignitions, new in the sense of coming from a source which only appeared when the triple ignitions were started. The processing of such new input volleys is described in the next few drawings.

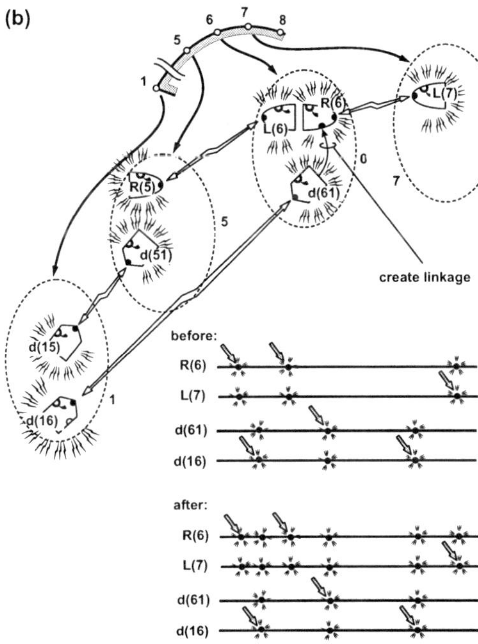

Fig. 20.2(b) Extending a long link while remaining on the contour (Cont.). All 1–6 co-ignitions are turned into triple (1–6–7) co-ignitions. At the end of the previous cycle (before the events in this series), the cells R(6) detect that their local conjugate cells, L(6), have resumed their passive following of all volleys of their facing conjugate cells, R(5). The resumption of passive following indicates that the 1–6 link has been completed (the stage is comparable to step (n) below). The completion of the 1–6 linkup serves as the signal for starting the next (1–7) linkup, the subject of the present sequence. As the first step of the new linkup, the synapse set in the cells R(6) marked out in the previous linkup (see steps (d) and (m) below) is promoted to "strong" status (blackened in the drawing), which makes them cause the postsynaptic cells to follow, passively, all volleys coming to them from d(61). The result is that every co-ignition between nodes 1 and 6 now becomes a triple ignition where nodes 1, 6, and 7 all ignite together. Through these triple ignitions, nodes 1 and 7 tell all cells reached both by nodes 1 and 7 that they are on the same contour

One additional step appears in the upper right corner, where the contour bridgehead R(7) is prepared for the next cycle, the linkup between nodes 1 and 8. It is the first preparation of the R bridgehead in the phase of the next cycle comparable to step (b) above. Without such preparation there is ambiguity in step (b) as to which is the "opposite-pointing" contour bridgehead. The cue for this action, like that for the passive following by the other cells, is the doubling of the rate of volleys. It is in this step, and its companion step in drawing (m), initializing the fulcrum of the next triangle-closing step, where it is utilized that the linkup proceeds in a chain-like sequence, and there is no uncertainty as to the choice of triangle to be closed next; the next triangle is simply determined by the current one.

It is noted that the volleys are only approximately simultaneous, because of conduction delays, and it is the role of the myelination of Gennari axons to make these

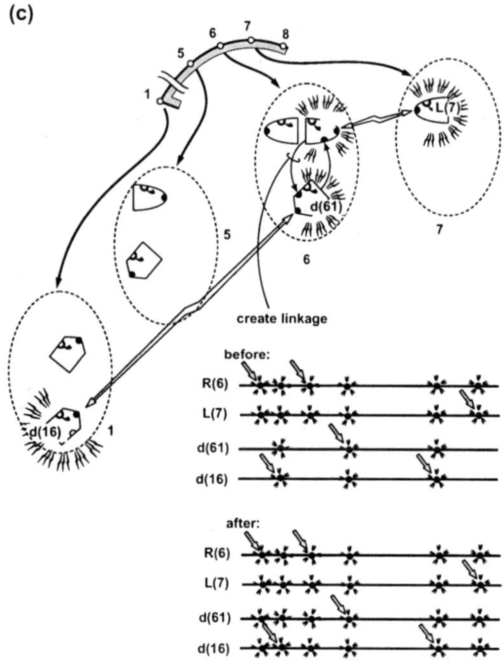

Fig. 20.2(c) Extending a long link while remaining on the contour (Cont.). All 6–7 co-ignitions are also turned into triple (1–6–7) co-ignitions. Cells d(61) detect that the volleys from R(6) follow their own volleys, and first mark their synapses, then promote them to strong synapses. Thereafter cells d(61) passively follow all volleys from R(6). The passive following, as always, is subject to the refractory time which prevents the ignitions from bouncing back and forth between nodes. The result is that now all 6–7 co-ignitions are also turned into 1–6–7 co-ignitions, making the triple ignitions noticeably more frequent than either of the component ignitions have previously been

delays small. The cross-correlograms published by Ts'o et al. (1986) suggest that the delays between ignitions 1–2 mm apart often reach 5 ms or more. The delays are not shown in the ignition sequences, for simplicity; only the *prime mover* of each co-ignition is indicated with its slanted arrow.

The cells reached from node 7 may or may not be also reached from node 6, therefore the input to d(17) from node 6 is shown with a broken line. If it exists, it is classified among the "familiar" inputs, the ones whose synapses are already marked before the onset of triple ignitions.

The term "introduction" is useful in this context (see (f), (g)), as used in sentences like "cells A *introduce* cells B to cells C" (and may be indicated in a shorthand notation as C:A→B). In the present case the introduction is of the form d(17):d(16)→d(71).

The introduction is accomplished through the repeatedly time-locked occurrence of volleys from the introduced cells and introducing cells, indicating that the two groups of cells "already know each other." In the present case the introduction also

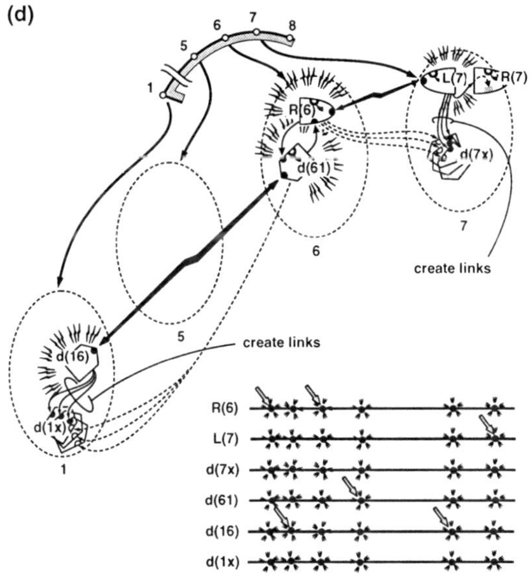

Fig. 20.2(d) Extending a long link while remaining on the contour (Cont.). Nodes 1 and 7 start emitting omnidirectional volleys. All the direction-coded cell groups in nodes 1 and 7, indicated as d(1x) and d(7x), detect that a linkup is desired of them, by the fact that the volleys from the local groups d(16) and L(7), respectively, are arriving to them roughly twice as often as before. The drawing implies that the "omnidirectional" volleys do not really go out in all directions, only in the most likely directions, but this assumption is not utilized

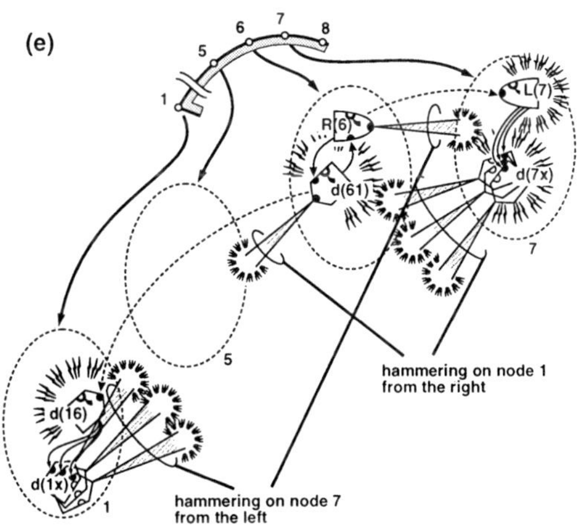

Fig. 20.2(e) Extending a long link while remaining on the contour (Cont.). Omnidirectional volleys create hammering from left and right. This drawing is included to illustrate the various sources of coordinated directed volleys at each triple ignition. The volleys constitute a "hammering" action which has the effect of telling all its recipient cells that all its sources are on the same contour (comparable to Fig. 21.1d)

(f)

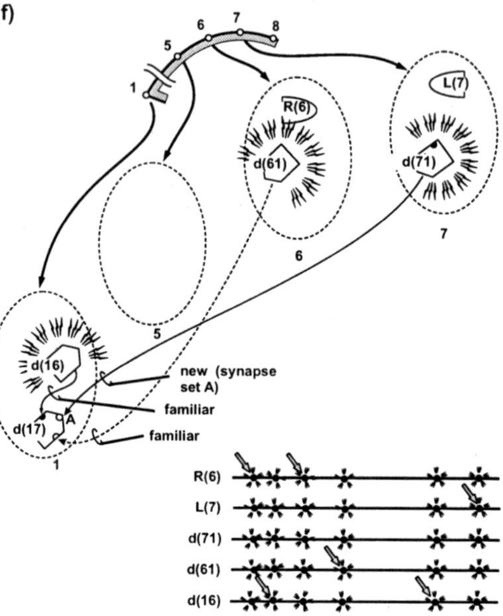

Fig. 20.2(f) Extending a long link while remaining on the contour (Cont.). Node 1 identifies a new source of volleys on the same contour. During the triple ignitions, if the volleys from node 7 reach any of the cells in a direction-coded cell group in node 1, the reached cells know, on the basis of the simultaneous arrival of volleys, that the source of the volleys is on the same contour which carries node 1. The implied message of simultaneity in the present context is that simultaneously arriving volleys originate on the same contour. The comments attached to loops around arrows here and below refer to things known or processed in the postsynaptic (arrowhead end) cells

(g)

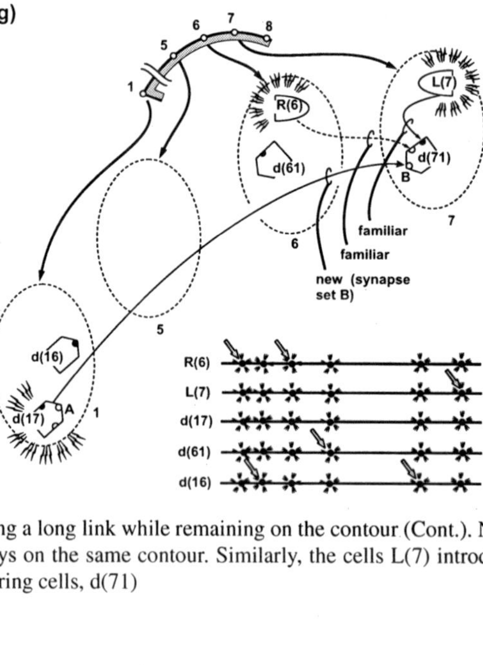

Fig. 20.2(g) Extending a long link while remaining on the contour (Cont.). Node 7 also identifies a new source of volleys on the same contour. Similarly, the cells L(7) introduce the remote cells d(17) to their neighboring cells, d(71)

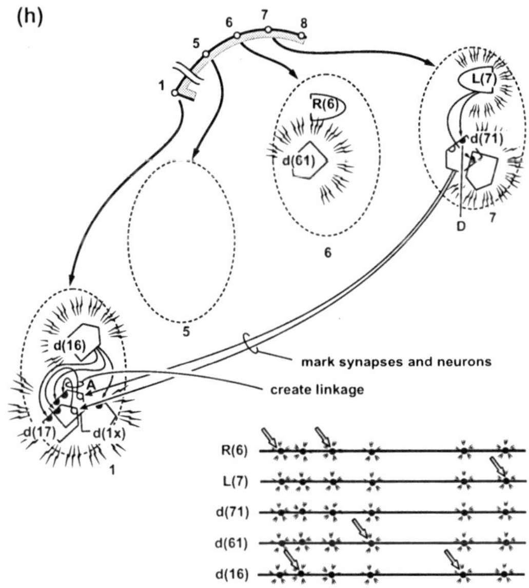

Fig. 20.2(h) Extending a long link while remaining on the contour (Cont.). Cells in node 1 reached from node 7 begin forming a group. At this point some cells in node 1 (through the synapse set designated A in the drawing) know that they are reached from node 7. The triple ignitions offer the cells a way to identify inputs from another node on the same contour; once that detection is made the cells reached by the inputs start forming an ignitable group

tells the cells d(17) that the cells newly introduced to them are on the same contour as they are themselves.

As mentioned in Sect. 20.1.8, forming an ignitable group on the basis of a new input is a challenging task because the cells in question only know that they themselves are reached from a new source; they do not know which other cells are reached from that source, therefore they do not know which synapses they must reinforce in order to create an ignitable group made up solely of cells reached from node 7. Also, as seen in the ignition sequences, all volleys, locally and remotely originating, are in sync. There are no out-of-sync volleys available to which the new group could tentatively synchronize itself and then link up with other out-of-sync cells.

And this is only the first part of the problem. The other part (Sect. 20.1.9) is that a group of cells needs to reach a certain critical size in order to be ignitable, and the number of cells reached from node 7 (which determines the size of the group of interest) is at first expected to be too small (especially when the nodes are far apart). It will then grow as the nodes grow (Fig. 17.1) until reaching is achieved, but the linkup protocols must be flexible enough to accommodate the uncertainty.

The solution chosen here (Sect. 20.1.8) relies on the assumption that direction-coded cells convey angle of relative direction by means of labeled lines; in other words different cells fire to signify different values of the angle. This in turn means that a direction-coded cell is in a position to know when another direction-coded

(i)

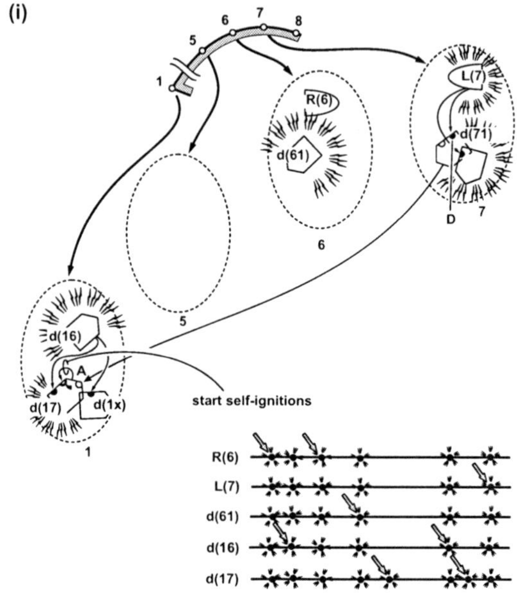

Fig. 20.2(i) Extending a long link while remaining on the contour (Cont.). The cells reached from the opposing satellite node start self-ignitions. As more of the same-pointing cells are reached from node 7, and start following other same-pointing cells, a stage is reached when these cells can form an ignitable group. When that happens, they start self-ignitions. As seen in the ignition sequences, this is the first drawing where there are any ignitions present out of sync with the triple ignitions (see marking arrows in trace d(17). At this point all cells still follow the triple ignitions

cell, whose firing reaches it, carries the same angle of relative direction as it does itself (see (h)).

Accordingly, the cells reached from node 7 reinforce the synapses they receive from all other direction-coded cells pointing the same way (shown in the drawing by parallel-shifted cell group icons).

The similarly pointing cells are a superset of the desired set, as the connections are made partly blindly. However, a few steps later, after active linkage is established and facing nodes are dynamically tracked by one another, the protocol becomes self-regulating (Chapter 16) and the extra connections are dropped.

As pointed out in Sect. 20.1.10, the cells of node 1 reached from node 7, as soon as they are numerous enough to support self-ignitions, know that enough of node 1 is reached to form a bridgehead. When they also detect that the input stream from node 7 contains independent self-ignitions, they know that the reaching is bidirectional. As a result they stop following the volleys from d(16) and with them the triple ignitions.

The "off-axis" cells of node 1, the ones not reached from node 7, look to the others for their cue to disconnect. When they detect that some local cells started emitting volleys out of sync with the triple ignitions, they know that node 7 reached node 1, but they don't know yet whether the reaching is bidirectional. In case it is

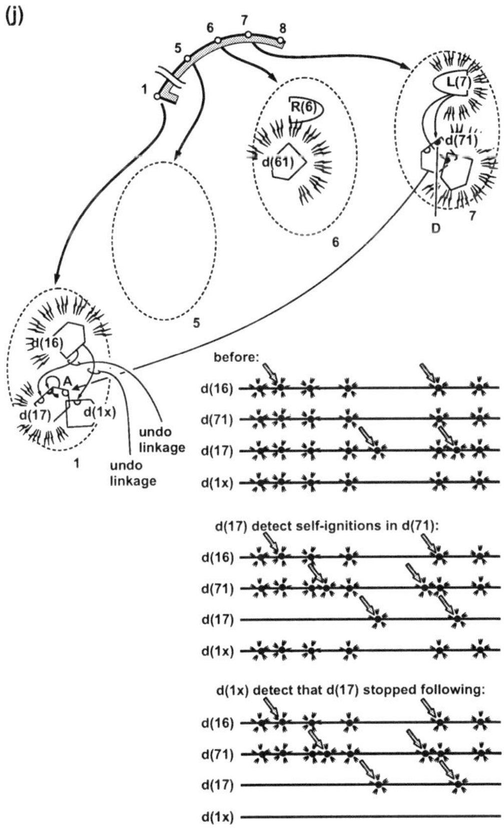

Fig. 20.2(j) Extending a long link while remaining on the contour (Cont.). Node 1 stops following the triple ignitions. After the node 1 cells reached from node 7 have started self-ignitions, they stop following the volleys from d(16), in other words they become independent of the triple ignitions. At the same time, the cells not reached from node 7 detect the presence of the out-of-sync volleys coming from their neighbors, and take them as the cue for them also to stop following the triple ignitions. The latter cells now fall silent

not, it may be preferable to continue emitting omnidirectional volleys and let the nodes grow some more. When they detect that the cells emitting the out-of-sync volleys have stopped following the triple ignitions, they stop following them too. The volleys coming from node 1 are now no longer omnidirectional.

The next task (also added to drawing (k)) is to tell the fulcrum somehow that there is no longer any need for the triple ignitions. I proceed here on the assumption (Sect. 20.1.13) that the signaling is achieved through the concept that once it is no longer necessary to passively follow a set of volleys, because the intended purpose of passive following has been achieved, the passive following can be stopped, which in turn tells the neurons which are no longer being followed that they can also tell their sources to disassemble that equipment devoted to the same task.

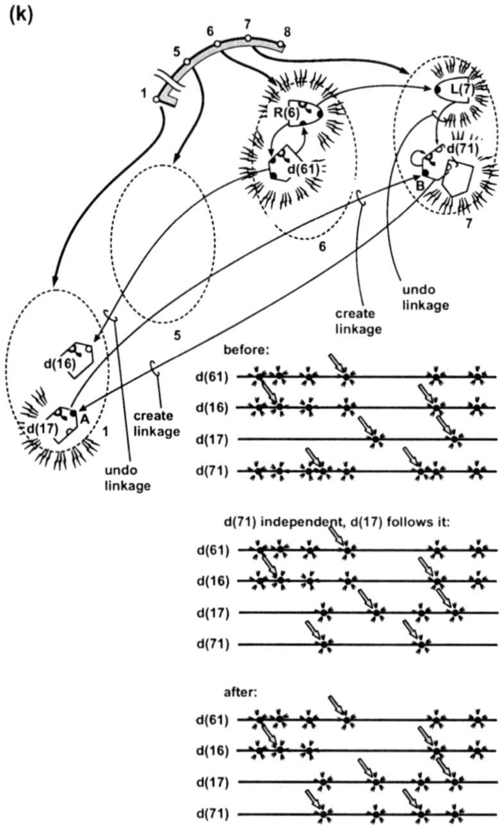

Fig. 20.2(k) Extending a long link while remaining on the contour (Cont.). Nodes 1 and 7 set up mutual excitation. At this point the mutually reaching bridgeheads in 1 and 7 ignite independently; next they form a strong bidirectional excitatory link, so an ignition by either one is passively followed by the other one (reverberation is prevented by the refractory period). From this time on, 1 and 7 are connected by an active link as described in Chapter 16

Accordingly, after the new bridgehead in node 1 is successfully linked up to one in node 7 and stops following the cells d(16), then subsequently the off-axis groups in node 1 stop following the cells d(16), the cells d(16) know that a task they mediated (whatever it was) has been accomplished, and briefly stop following the cells d(61), which are their source of the triple ignitions. The intent is thereby to tell d(61) to also undo the passive following of volleys from R(6), because their purpose has also been achieved.

After the cells d(16) stop following the cells d(61), the cells d(61) respond by stopping to follow the cells R(6), and clearing whatever chemical marker told them that a linkup was in progress. Similarly, the corresponding things happen on the node 7 side in the node's own time sequence. After the cells L(7) stop following R(6), the

(l)

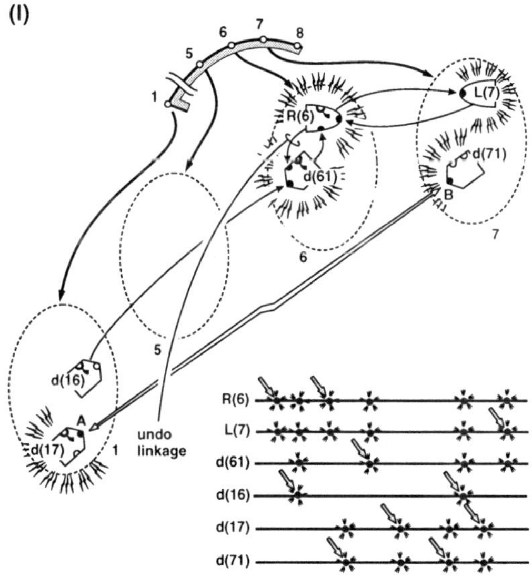

Fig. 20.2(l) Extending a long link while remaining on the contour (Cont.). The fulcrum bridgehead at the node 1 side disengages itself. The bridgehead d(61) in the fulcrum detects that it is not being followed by the node 1 bridgehead, and takes this as an instruction to break itself away from R(6), as part of stopping the triple ignitions, because node 1 has completed its side of the desired linkup with node 7

(m)

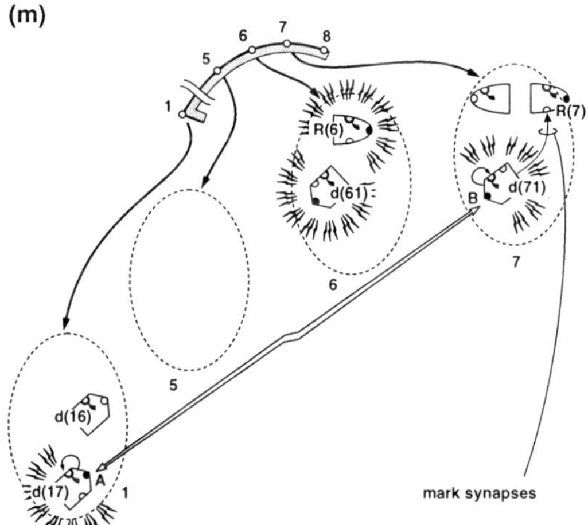

Fig. 20.2(m) Extending a long link while remaining on the contour (Cont.). Node 7 is prepared for its role as fulcrum of the next linkup. Next, node 7 marks the synapses from the synapse pool prepared in step (d), at the upper right of the drawing. This marking is in preparation for step (b) in the next linkup, the one which is to take place between nodes 1 and 8

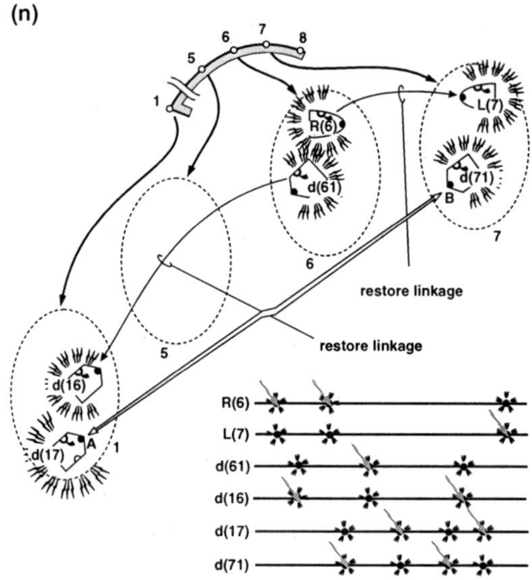

Fig. 20.2(n) Extending a long link while remaining on the contour (Cont.). Links 1–6 and 6–7 are restored. Cell groups d(16) and L(7) detect that the rate of ignitions coming from their facing conjugate groups in the fulcrum node, d(61) and R(6), has fallen back down by about half, and resume their passive response to d(61) and R(6). Since the self-ignitions of d(16) and L(7) have never been stopped, and the strong synapses which earlier supported passive following of node 1 and 7 by d(61) and R(6) are still in place, the last change restores all links to their original operation

cells R(6) respond by stopping to follow d(61), and by clearing the chemical marker which told them that a linkup was in progress.

The result is that d(61) and R(6) now ignite independently, and consequently on average about half as often as they did while passively following each other.

The cells L(7) and the cells d(16) detect the reduction of the rate of ignitions, interpret it as indicating the successful completion of linkup, and resume following of their facing groups. The original linkage between nodes 1 & 6, and nodes 6 & 7 is restored.

Chapter 21
Completing a Triangle of Links

Triangles are special in two-dimensional polygon graphics (as are tetrahedrons in 3D polygon graphics). As was mentioned, the higher cortex cannot distinguish a square from a rectangle on the basis of the relative directions between adjacent corners alone; it needs at least one diagonal, because the shape of quadrangles and higher polygons is indeterminate when only their angles are specified, but the shape of triangles isn't. This makes it important to close as many triangles as possible among the grid of nodes, with relative direction links.

In the sequential linkup to a corner, each step is a special case of triangle closing, where two corners are on an existing contour link and the third is a fixed corner point. The next sequence describes the closing of triangles made up completely of relative direction links. Fulcrum nodes initiating the sequence are chosen without any regularity.

In the description chosen here, the closing of triangles takes the form of a firing game; it goes on all the time and chooses its targets in many places independently of one another, wherever the process detects a pair of links forming an open triangle which can be closed.

The expected result is that after a while enough triangles are closed to determine the shapes uniquely and in fact to over-determine them, providing a kind of redundancy which is usual and desirable in the brain context.

21.1 Closing a Triangle

The closing of a triangle of links amounts to the elementary "deduction" which utilizes the transitive property of the "same contour" relation, a relation reduced here to hardware by means of Gennari linkage.

The triangle closing sequence is in all its essentials identical to the sequence of Figs. 20.1 and 20.2, because those are merely special cases of triangle closing sequences.

One slight difference is that in the chain linkup drawings of Chapter 20 one of the corner points of the originally open triangle was always the same node, node 1, and the side opposite that node was taken to be a *contour* link, whereas in the triangle

C.R. Legéndy, *Circuits in the Brain*, DOI 10.1007/978-0-387-88849-1_21,

closing linkup any of the corners can be anywhere, so the fulcrum node will simply be called "B" (as in Sec. 20.1), and the satellite nodes "A" and "C," and all sides of the triangle are assumed to be *directional* links (Figs. 21.1(a–c) and Fig. 21.2(a–d)).

Another slight difference is that in the chain linkup the "omnidirectional volleys," aiming to reach a node of unknown location, can sometimes be confined to

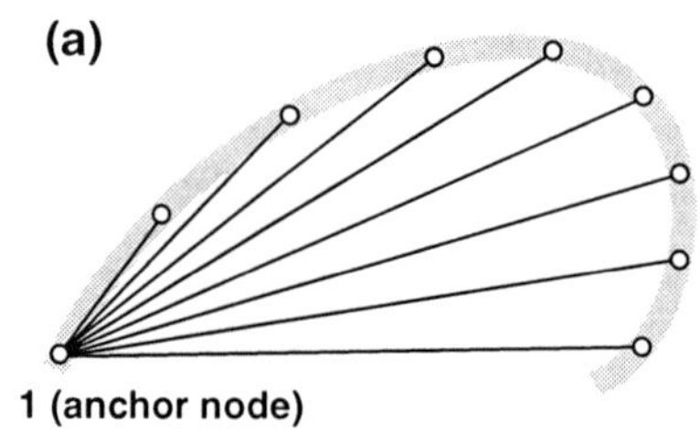

Fig. 21.1(a) Completing a triangle of links. Links as they look at the completion of chain linkup. At the end of the chain linkup, in other words after the steps of Figs. 20.1 and 20.2 are repeated until the next corner node is reached, the contour has long links fanning out from a single shared node, the original anchor node, to all other nodes

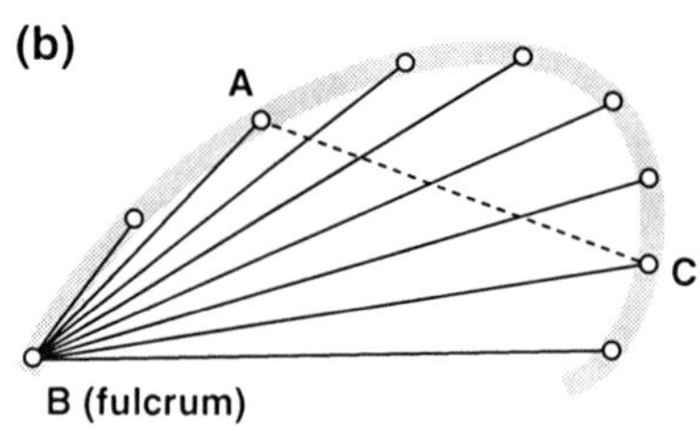

Fig. 21.1(b) Completing a triangle of links (Cont.). At first, the former anchor node is the only possible fulcrum node. Since at first the only node shared between any two long links is the original anchor node, and since the fulcrum of a linkup must be a node shared between the already existing links of the triangle, the fulcrum of the first linkups can only be the original anchor node

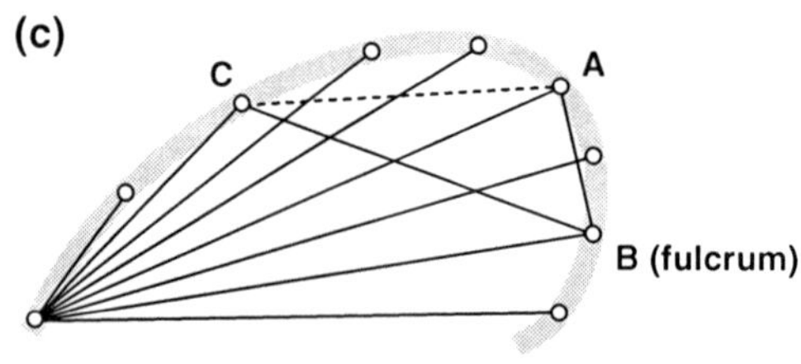

Fig. 21.1(c) Completing a triangle of links (Cont.). After a while, any node may be detected as a possible fulcrum. Once some of the long links meet in nodes other than the original anchor node, the nodes at the meeting points can serve as fulcrum. The catch is that after a while many triangles will be closed, and good protocol demands that before a node can initiate a linkup, and serve as fulcrum of the linkup, it must know more than just that it is at the meeting point of two links. It must also know that the far ends of the links are not yet interconnected; in other words that the triangle is still open. This is the issue addressed next, in Fig. 21.2

a limited range of angles around the volleys aimed at the already familiar fulcrum node, because one side of the triangle being closed is often small, but in the general triangle closing protocol these volleys have to be really *omnidirectional*, because none of the sides of the triangle can be assumed to be short.

21.2 How to Spot "Open" Triangles: The Three-Element Problem

In every sequence the first question that may be asked is how the neurons know when to go into the sequence? In the sequential linkup of Figs. 20.1 and 20.2 the answer is that the natural progression of the sequence tells the nodes to commence the linkup. The fulcrum knows that it just completed a linkup, starts the triple ignitions, and thereby initiates the next linkup process.

In the general triangle closing sequence, there is no corresponding progression. Since in a developing network there are, at any given time, expected to exist some open triangles and some closed ones, triangle closing is expected to be an ongoing process. The question arises, how does a node know that it is properly located to be the fulcrum node of a triangle closing sequence? Another way to put it: how a node containing two bridgeheads is supposed to know that the nodes at the *other ends* of the bridgeheads are not yet tied together?

It is a tricky issue, and brings up a more general one: the "three-element problem." The bridgehead on node B facing toward node A knows, through its linkup dialog, that node A has a bridgehead toward node B, but it does not know about other bridgeheads node A may have. And yet, if node B is to serve as fulcrum for a successful triangle closing sequence, it needs to know that node A carries a link to some third node C, and even needs to know that it (node B) carries a bridgehead toward that very same node C.

In other words we need to be able to plant knowledge in one node concerning a relation between *two other nodes*.

The solution described here (Fig. 21.2(a–d)) is to place "marking labels" on those synapse sets which would be changed as the first step of any triangle closing sequence (corresponding to steps (b) and (c) in Fig. 20.2), the step where two bridgeheads of a node are temporarily fused together, turning the node into the fulcrum of a triangle closing sequence, and initiating the necessary triple ignitions. The labels say, in effect, not to bother fusing these bridgeheads and creating a fulcrum for a new triangle closing process, because the triangle that would result already exists.

The trick to putting these labels in place is to add steps to every triangle closing sequence (including the sequence of Fig. 20.2, which also amounts to triangle closing), and in the course of the sequence add six labels altogether, two for each corner of the triangle, for the synapses going both ways between the bridgeheads serving the corner.

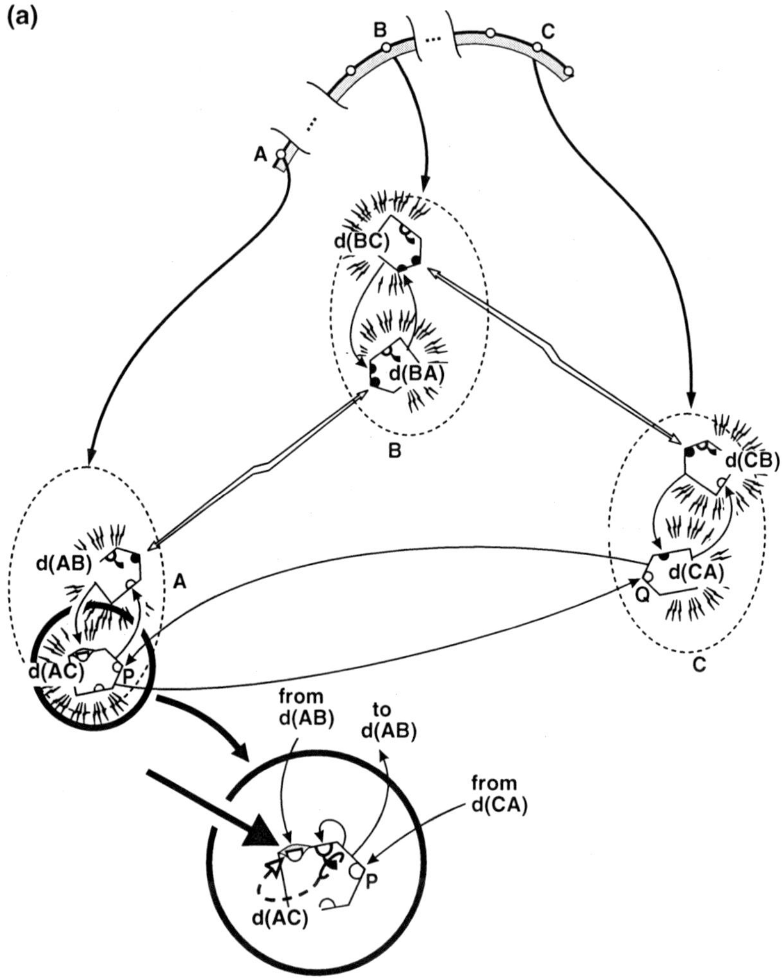

Fig. 21.2(a) Labeling the corners of a triangle ABC, when closing it. Labeling new bridgehead at A after it becomes independent. The nodes, as in Fig. 21.1, are denoted with letters A, B, C, and magnified versions of the nodes of interest are included where needed. Since the steps of general triangle closing and the steps of Fig. 20.2 are the same, reference will repeatedly be made to Fig. 20.2. At the first piece of evidence of successful triangle closing (at the A side of the triangle), the start of independent self-ignitions by the cells d(AC), the synapses from d(AB) to d(AC) are tentatively labeled

The idea is that there are brief intervals of time during triangle closing for each of the six locations, during which the successful linkup of the other two nodes is known at those locations. The labels can be created during those intervals. Once in place, the labels can act as source of the same knowledge at later times, when the knowledge is no longer obtainable through signal transmission.

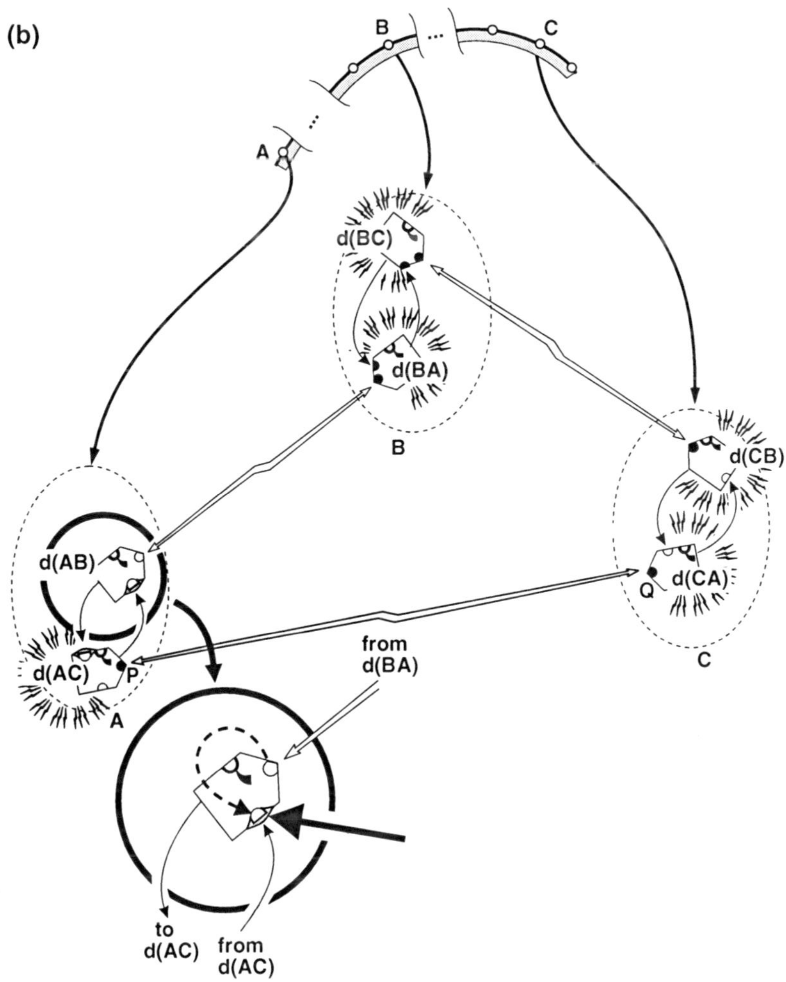

Fig. 21.2(b) Labeling the corners of a triangle ABC, when closing it (Cont.). Labeling old bridgehead at A after the new one stops following it. The next synapses to be labeled (still concentrating on the A side of the triangle) are the ones inside node A going in the opposite direction, from d(AC) to d(AB). They are the synapses from the nascent bridgehead to one of the old bridgeheads, and their host neurons know of the fact that the nascent bridgehead stopped following the triple ignitions, meaning that it is now an ignitable group solidly reached from node C

It must be noted that one bridgehead can participate in many triangles, but a pair of bridgeheads can only participate in one; therefore when the mutual interfaces of the bridgeheads within each node are labeled, the labeling singles out one triangle. By the same token, in order to label a pairing of bridgeheads, it is necessary to tell each bridgehead about the other that it should not combine with, through some chemical marking.

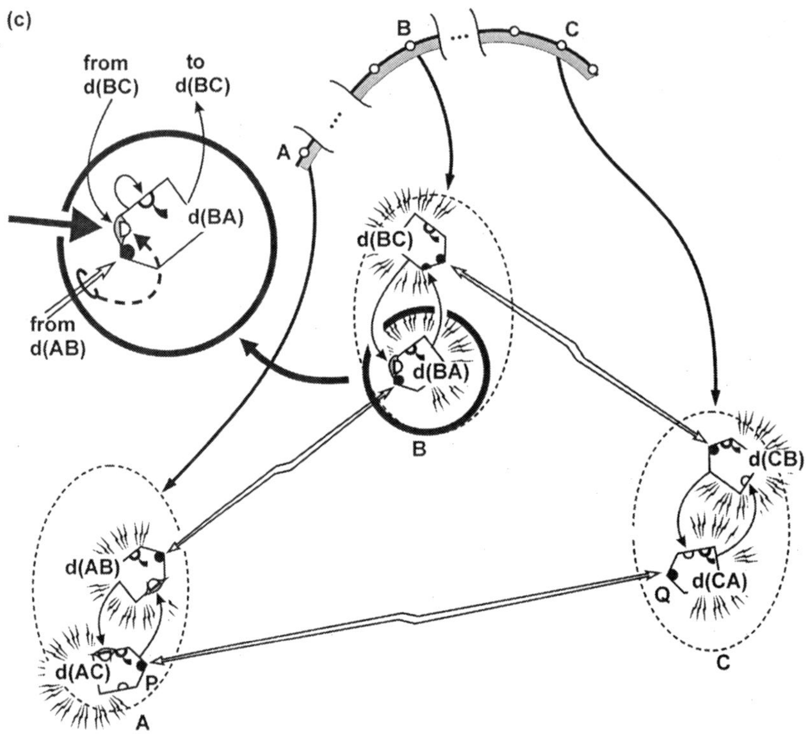

Fig. 21.2(c) Labeling the corners of a triangle ABC, when closing it (Cont.). Labeling old bridgehead at B after bridgehead at A stops following it. The last labeled synapse set (on the node A side of the triangle) is the d(BC)→d(BA) synapse set, labeled when node A tells node B about its linkup with C, by stopping to follow its ignitions. Here, too, a confirmation is necessary, in the form of assurance that d(BC) has also successfully gone through the steps of its dialog with d(CB) in the form of cessation and resumption of following

The labeling in question, as will be noted, is not made on the "business end" of the bridgeheads, in other words not on the synapses at the interface between two facing nodes, but always on synapses connecting two bridgeheads of *the same* node.

The first indication anywhere among the nodes A, B, C that a triangle may be successfully closed is when one of the satellite nodes confirms the independent volleys coming from the other, and starts self-ignitions (Fig. 20.2(i)).

At this point, node A proceeds to disconnect itself from the triple ignitions (Fig. 20.2(j)). The latter act of disconnecting is the signal for making the labeling on the synapse set which was previously blackened (for passive-follow operation) and is subsequently not blackened.

At the time in question the neurons carrying synapses know that they are reached by their opposing node (node 7 in Fig. 20.2 and node C in the present drawing). The act of disconnecting is the result of repeated volleys received from node C, and amounts to the first phase of linkup between A and C.

(d)

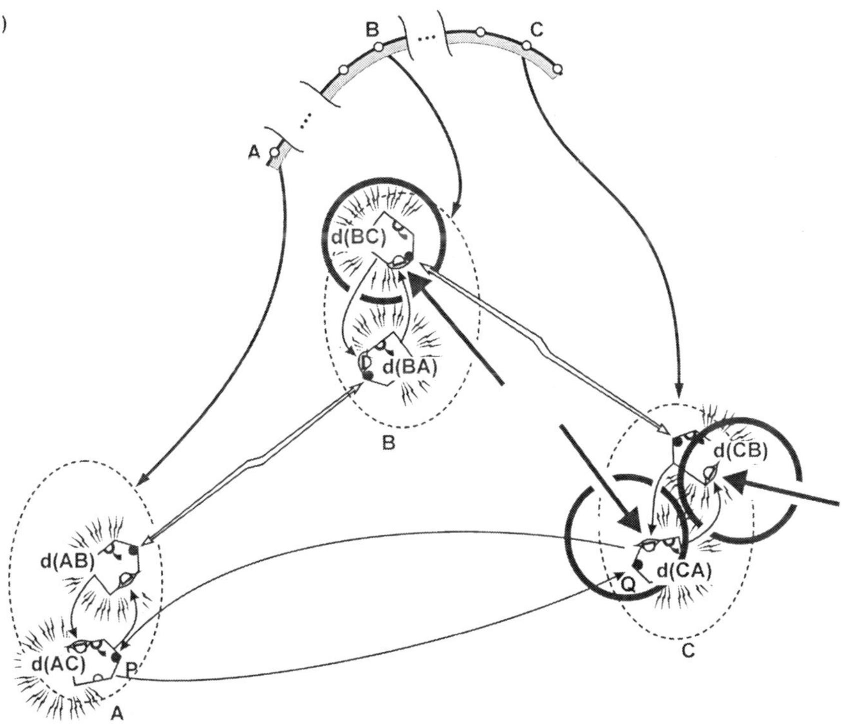

Fig. 21.2(d) Labeling the corners of a triangle ABC, when closing it. Placing the same three labels at the C-side of the triangle. The corresponding steps at the node C side of the triangle proceed on their own, and independently, sometimes ahead of the steps at the node A side, sometimes behind. This last drawing summarizes C-side versions of the steps (a), (b), and (c)

The special labeling is indicated in the present drawings by a grayed "blister" drawn over each labeled synapse set (the present drawing concentrates on node A, leaving the corresponding events in node C to drawing (d)). The first synapse set so labeled is, in other words, the set of synapses in node A from the bridgehead d(AB) to the nascent bridgehead d(AC). The labeling is created and confirmed in the process of linkup confirmation as nodes A and C proceed in their linkup dialog, in which self-ignitions are locally distinguishable from remotely initiated ignitions.

The labeling of d(AB)→d(AC) synapses is at this point only tentative, because it is not yet certain that the other parts of the triangle closing sequence will go through without failure. The confirmation step is necessary (as is a similar confirmation step to the labeling in drawings (b) and (c) below). In this case the confirmation is that the d(CA)→d(AC) volleys become random and independent of the ignitions arriving to d(AC) from d(AB), and occur at a decreased frequency, indicating that the ignitions of d(CA) have also become independent of the triple ignitions. If the confirmation is not made, the labeling is rolled back after a while.

It will be noted that the confirmation step, like the labeling itself, is designed to only require information available to the neurons d(AC), the ones which receive the labels.

It is necessary to emphasize that the set of neurons making up the bridgehead carrying the initially labeled synapses is a dynamically changing set, as is the labeled synapse set on the neurons making up the bridgehead, since the marking is understood to be continually updated in the course of tracking as the bridgeheads change in the course of retinal drift (*dynamic* marking). This updating proceeds at the same time as refining of the nascent synapse sets. Since the labeling is assumed to be made (by some chemical means) a *property* of the synapse set, it will, by the same assumption, be passed along from neuron to neuron and synapse to synapse, as refinement and dynamic marking continually update the synapse set.

The d(AC)→(AB) labeling is subject to a confirmation step, which fails to occur if something goes wrong at the other side (the C-node side) of the process. When, in node A, the new bridgehead, d(AC), stops following the old one, d(AB), the corresponding node B bridgehead, d(BA), is supposed to stop following d(AB). Then after a while, when the fulcrum node is disassembled, d(BA) is supposed to resume following d(AB). This latter resumption of following (detectable at d(AB)) is the step which confirms the labeling from d(AC) to d(AB), and tells it that the triangle closing is successfully completed.

Chapter 22
All-to-All Linkup on Smaller Shapes, Utilizing Chain Ignitions

22.1 Indiscriminate Linkup of All Nodes

When a shape is simple and relatively small, it does not require the step-by-step linkup described in Fig. 20.2 and then separate steps of closing more triangles. The rationale for the sequential process of 20-2 and the subsequent individual steps of triangle closing is that it can accommodate contours with many nodes. When a contour has many nodes, creating a link between every pair of nodes is wasteful; for instance from 20 nodes one can form 190 node pairs, which are a lot more than necessary to dissect the shape into triangles.

The sequential linkup, combined with individual steps of triangle closing, saves time, because only the Fig. 20.2 steps need to be performed sequentially; a subsequent batch of a few dozen triangle closing steps can proceed mostly in parallel without being overwhelmed by crosstalk.

The same problems do not arise when a shape has for instance only five nodes. Then (if the contour links are already in place) it is feasible to arrange a much faster process, as shown in Fig. 22.1(a)–(e), with a sequence of spreading multi-node co-ignitions, in which all direction-coded cells ignite at the same time and link up if they can reach each other.

(a)

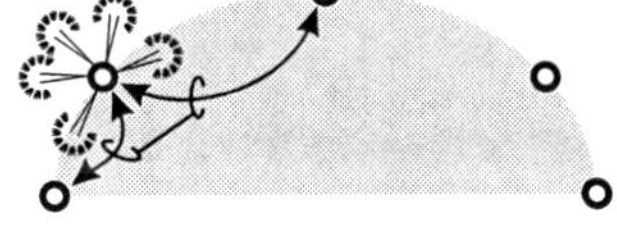

Fig. 22.1(a) The chain ignition game useful in smaller shapes. A node starts chain ignitions: all its Gennari cells fire together. A node detects that it has both contour links but no relative direction links, and goes into "chain ignition" mode. It creates a two-way link between its contour bridgeheads, and lets its contour ignitions spread to all its direction-coded cells with the result that all its contour cells and all its direction-coded cells fire every time either of its contour bridgeheads self-ignites (comparable to Fig. 20.2 (b, c, d))

C.R. Legéndy, *Circuits in the Brain*, DOI 10.1007/978-0-387-88849-1_22,

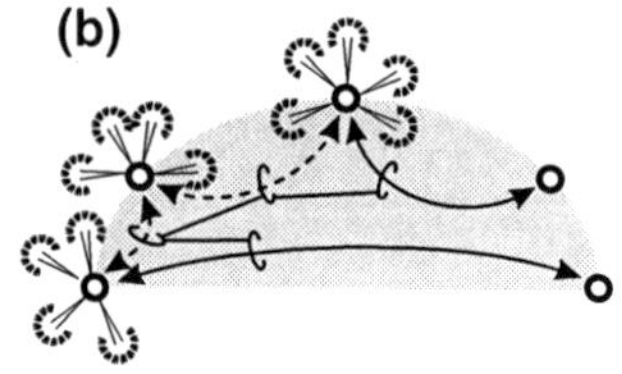

Fig. 22.1(b) The chain ignition game useful in smaller shapes (Cont.). The chain ignition spreads via contour links to next nodes. The two nodes reached by contour links from the chain-igniting node, similar to the first node, also, independently, go into chain ignition mode, interlinking their contour and direction-coded cells, so now all these cells in all three nodes give off omnidirectional volleys in unison with the self-ignitions of either of them (no comparable step in Fig. 20.2). When chain ignitions start from two or more centers independently, the igniting sets coalesce. They detect foreign chain ignitions reaching them by their increased rate of volleys and by the synchrony between their contour volleys and direction-coded volleys, then they bring their ignitions into synchrony with the other chain ignitions

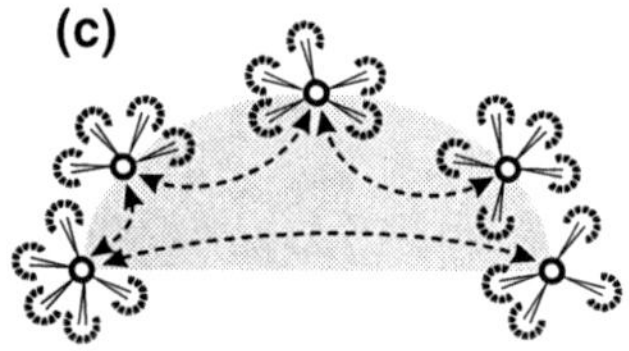

Fig. 22.1(c) The chain ignition game useful in smaller shapes (Cont.). If spreading stops soon enough, cells mark their in-sync inputs. The spreading process continues. As more and more nodes become active, the direction-coded cells in them receive in-sync volleys from direction-coded cells of more and more nodes. Since the shape shown does not have very many nodes, the spreading soon stops and the input volleys level off. The cells which detect that their input volleys have leveled off into a repeating pattern before noise becomes overwhelming, mark their synapse sets bringing new input (comparable to 20-2(f, g)) and the cells that do not, stop following the mass-ignitions. If the shape has too many nodes, and the volleys do not level off but keep growing until they become more than the nodes can handle, the nodes exit the chain ignition mode and restore their cells

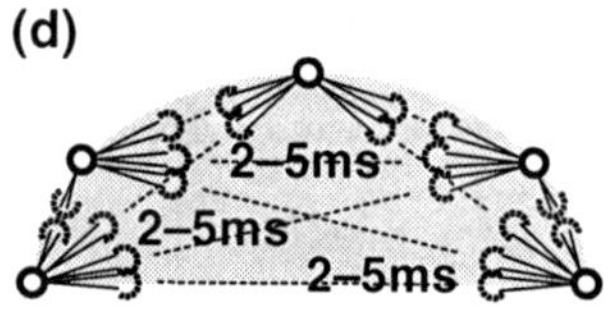

Fig. 22.1(d) The chain ignition game useful in smaller shapes (Cont.). Cells with marked synapses start self-ignitions. The cells which have marked their synapses transition to an intermediate mode mark their synapses from similar direction-coded cells of the same node, for the purposes of self-ignition support (comparable to Fig. 20.2(h)). Next they start self-ignitions (comparable to Fig. 20.2(i)) and stop following the local contour cells (comparable to Fig. 20.2(j)), and start passively following neurons generating any out-of-sync volleys from other nodes (comparable to Fig. 20.2(k)). The nodes do not need to await being reached, as in Fig. 20.2(j), by all remote nodes

(e)

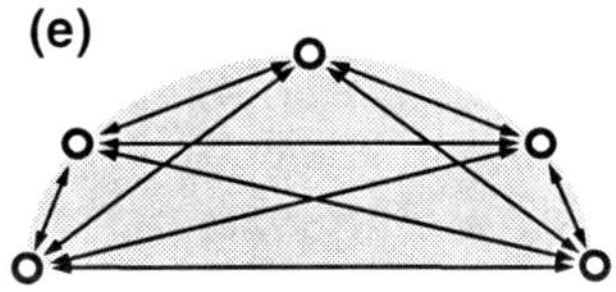

Fig. 22.1(e) The chain ignition game useful in smaller shapes. Contour cells detect not being followed and exit chain mode. At this point all nodes are linked to all other nodes through active relative direction links. The contour cells detect that they are no longer being followed by the direction-coded cells, and disconnect themselves from the other local contour bridgehead (comparable to Fig. 20.2(l)), then exit the chain ignition mode (the direction-coded cells already exited it)

Closing Remarks

It has been my purpose throughout this book to make the things I presented appear self-evident. If somebody looks at my drawings and complains that their message is too obvious to be published, I will consider my efforts successful.

At the time of this writing I have, of course, no idea how much of what I write here will prove correct.

I am very confident that most of my methodology will survive: the idea of reaching, the idea of surprise, the principle of overwhelming odds, the logistic considerations dealing with local knowledge, and others. I am optimistic, in general, since my assumptions are not outlandish. After the dust settles, and after what I wrote here is re-worked to square with the facts, I assume the theoretical methods I have sketched out here will still be around.

It has cast a shadow over the development of neuroscience that the massively productive two-way interaction between theory and experiment, which is the engine driving the success of physics, chemistry, and molecular biology, is all but absent in brain research. But there is no good reason why this should be so; and if my book can improve the interaction, even a little bit, its goals will have been achieved.

References

Abeles, M. (1982) Role of the cortical neuron: Integrator or coincidence detector? Isr. J. Med. Sci. vol 18, pp 83–92

Ahmed, B., Anderson, J. C., Douglas, R. J., Martin, K. A. C., and Whitteridge, D. (1993) A method of estimating net somatic input current from the action potential discharge of neurons in the visual cortex of the anesthetized cat. J. Physiol. (Lond.) vol 459, p 134

Alverez, G., Giuditta, A., and Koenig, E. (2000) Protein synthesis in axons and terminals: Significance for maintenance, plasticity and regulation of phenotype. With a critique of slow transport theory. Prog. Neurobiol. vol 62, pp 1–62

Araque, A., Carmignoto, G., and Haydon, P. G. (2001) Dynamic signaling between astrocytes and neurons. Annu. Rev. Physiol. vol 63, pp 795–813

Arbib, M. A. (1964) Brains, Machines and Mathematics. (New York: McGraw-Hill)

Arbib, M. A. (Ed.) (2003) The Handbook of Brain Theory and Neural Networks, 2nd Ed. (MA: Cambridge and London: MIT Press)

Ariav, G., Polsky, A., and Schiller, J. (2003) Submillisecond precision of the input-output transformation function mediated by fast sodium dendritic spikes in basal dendrites of CA1 pyramidal neurons. J. Neuroscience. vol 23, pp 7750–7758

Ascher, P. and Nowak, L. (1988) The role of divalent cations in the N-methyl-D-aspartate responses of mouse central neurones in culture. J. Physiol. (Lond.) vol 399, pp 247–266

Baillarger, J. G. F. (1840) Recherches sur la structure de la couche corticale des circonvolutions du cerveau. Mém. Acad. roy. Méd. vol 8, pp 149–153

Bak, M., Girvin, J. P., Hambrecht, F. T., Kufta, C. V., Loeb, G. E., and Schmidt, E. M. (1990) Visual sensations produced by intracortical microstimulation of the human occipital cortex. Med. Biol. Eng. & Comput. vol 28, pp 257–259

Baker, B. J., Kosmidis, E. K., Vucinic, D., Falk, C. X., Cohen, L. B., Djurisic, M., and Zecevic, D. (2005) Imaging brain activity with voltage-and calcium-sensitive dyes. Cell. Mol. Neurobiol. vol 25, pp 245–282

Belmonte, C. and Viana, F. (2008) Molecular and cellular limits to somatosensory specificity. Mol. Pain vol 4, pp 14–17 (open access, doi:10.1186/1744-8069-4-14)

Beurle, R. L. (1956) Properties of a mass of cells capable of regenerating pulses. Phil. Trans. Roy. Soc. Lond., B, Biol. Sci. vol B240, pp 55–94

Binzegger, T., Douglas, R. J., and Martin, K. A. (2004) A quantitative map of the circuit of cat primary visual cortex. J. Neurosci. vol 24, pp 8441–8453.

Blinkov, S. M. and Glezer, I. I. (1968) The Human Brain in Figures and Tables. (New York: Plenum)

Bliss, T. V. P. and Collingridge, G. L. (1993) A synaptic model of memory: Long-term potentiation in the hippocampus. Nature vol 361 pp 31–39

Braitenberg, V. (1978) Cell assemblies in the cerebral cortex. In: Heim, R. and Palm, G. (Eds.) Theoretical Approaches to Complex Systems (Berlin: Springer) pp 171–188

Braitenberg, V. and Schüz, A. (1998) Cortex: Statistics and Geometry of Neuronal Connectivity, 2nd Ed. (Heidelberg: Springer-Verlag)

Brodmann, K. (1909) Vergleichende Lokalisationslehre der Grosshirnrinde in ihren Prinzipien dargestellt auf Grund des Zellenbaues. (Leipzig: Johann Ambrosius Barth Verlag)

Buzsáki, G. (2004) Large-scale recording of neuronal ensembles. Nature Neurosci vol 7, pp 446–451

Caianiello, E. R. (1961) Outline of a theory of thought-processes and thinking machines. J. Theor. Biol. vol 2, pp 204–235

Chino, Y. M., Smith, E. L., Kaas, J. H., Sasaki, Y., Cheng, H. (1995) Receptive field properties of deafferentated visual cortical neurons after topographic map reorganization in adult cats. J Neurosci vol 15, pp 2417–2433

Cox, D. R. and Smith, W. L. (1954) On the superposition of renewal processes. Biometrika vol 41, pp 91–99

Cragg, B. G. (1975) The density of synapses and neurons in normal, mentally defective and aging human brains. Brain vol 98, pp 81–90

Creutzfeldt, O. D., Garey, L. J., Kuroda, R., and Wolff, J.-R. (1977) The distribution of degenerating axons after small lesions in the intact and isolated visual cortex. Exp. Brain Res. vol 27, pp 419–440

Creutzfeldt, O. D., Lux, H. D., and Nacimiento, A. C. (1964) Intracelluläre Reizung corticaler Nervenzellen. Pflügers Arch. Gesamte Physiol. vol 281, pp 129–151

Degos, B., Deniau, J.-M., Thierry, A.-M., Glowinski, J., Pezard, L., and Maurice, N. (2005) Neuroleptic-induced catalepsy: Electrophysiological mechanisms of recovery induced by high-frequency stimulation of the subthalamic nucleus. J. Neurosci. vol 25, pp 7687–7696

Denk, W., Strickler, J., and Webb, W. (1990). Two-photon laser scanning fluorescence microscopy. Science vol 248, pp 73–76

Destexhe, A., Rudolph, M., Fellous, J.-M., and Sejnowski, T. J. (2001) Fluctuating synaptic conductances recreate in vivo-like activity in neocortical neurons. Neuroscience vol 107, pp 13–24

Djurisic, M., Antic, S., Chen, W. R., and Zecevic, D. 2004. Voltage imaging from dendrites of mitral cells: EPSP attenuation and spike trigger zones. J. Neurosci. vol 24, pp 6703–6714

Douglas, R. J., Martin, K. A., and Whitteridge, D. (1991) An intracellular analysis of the visual responses of neurones in cat visual cortex. J. Physiol. (Lond.) vol 440, pp 659–696

Gasparini, S. and Magee, J. C. (2006). State-dependent dendritic computation in hippocampal CA1 pyramidal neurons. J. Neurosci. vol 26, pp 2088–2100.

Ge, W. P. and Duan, S. M. (2007) Persistent enhancement of neuron-glia signaling mediated by increased extracellular K+ accompanying long-term synaptic potentiation. J. Neurophysiol vol 97, pp 2564–2569

Gennari, F. (1782) De peculiari structura cerebri nonnullisque eius morbis—Paucae Aliae Anatom. Observat. Accedunt. (Regio Typographeo, Parma, Italy).

Genoud, C., Quairiaux, C., Steiner, P., Harald Hirling, H., Welker, E., Graham, W., and Knott, G. W. (2006) Plasticity of astrocytic coverage and glutamate transporter expression in adult mouse cortex. Pub Library Sci. Biol. vol 4, pp 2057–2064

Gerstein, G. L. and Kiang, N. Y.-S. (1960) An approach to the quantitative analysis of electrophysiological data from single neurons. Biophysical J. vol 1, pp 15–28

Geschwind, N (1965) Disconnexion syndromes in animals and man. Brain vol 88, pp 237–294

Gilbert, C. D., Das, A., Ito, M., Kapadia, M., and Westheimer, G. (1996) Spatial integration and cortical dynamics. Proc. Natl. Acad. Sci. USA vol 93, pp 615–622

Gilbert, C. D. and Kelly, J. P. (1975) The projection of cells in different layers of the cat's visual cortex. J. Comp. Neurol. vol 163, pp 81–106

Gilbert, C. D. and Wiesel, T. N. (1979) Morphology and intracortical projections of functionally characterised neurones in the cat visual cortex. Nature vol 280, pp 120–125

Gilbert, C. D. and Wiesel, T. N. (1983) Clustered intrinsic connections in cat visual cortex. J. Neurosci. vol 3, pp 1116–1133

Gilbert, C. D. and Wiesel, T. N. (1990) The influence of contextual stimuli on the orientation selectivity of cells in the primary visual cortex of the cat. Vision Res. vol 30, pp 1689–1701

Giuditta, A., Chun, J. T., Eyman, M., Cefaliello, C., Bruno, A. P., and Crispino, M. (2008) Local gene expression in axons and nerve endings: The glia-neuron unit. Physiol. Rev. vol 88, pp 515–555

Gross, C. G. (2002) Genealogy of the "grandmother cell." Neuroscientist vol 8, pp 512–518

Gross, D. and Harris, C. M. (1998) Fundamentals of Queueing Theory 3rd Ed. (Canada: John Wiley and Sons, Ltd.)

Gross, C. G., Rocha-Miranda, G. E., and Bender, D. B. (1972) Visual properties of neurons in the inferotemporal cortex of the macaque. J Neurophysiol vol 35, pp 96–111

Hanes, D. P., Thompson, K. G., and Schall, J. D. (1995) Relationship of presaccadic activity in frontal eye field and supplementary eye field to saccade initiation in macaque: Poisson spike train analysis. Exp. Brain Res. vol 103, pp 85–96

Hebb, D. O. (1949) The Organization of Behavior. (New York: Wiley)

Holt, G. R., Softky, W. R., Koch, C., and Douglas, R. J. (1996) Comparison of discharge variability in vitro and in vivo in cat visual cortex neurons. J. Neurophysiol. vol 75, pp 1806–1814.

Hubel, D. H. and Wiesel, T. N. (1959) Receptive fields of the single neurones in the cat's striate cortex. J. Physiol. (Lond.) vol 148, pp 574–591

Hubel, D. H. and Wiesel, T. N. (1962) Receptive fields, binocular interaction and functional architecture in the cat's visual cortex. J. Physiol. (Lond.) vol 160, pp 106–154

Hubel, D. H. and Wiesel, T. N. (1968) Receptive fields and functional architecture of monkey striate cortex. J. Physiol. (Lond.) vol 195, pp 215–243

Larkum, M. E., Waters, J., Sakmann, B., and Helmchen, F. (2007) Dendritic Spikes in Apical Dendrites of Neocortical Layer 2/3 Pyramidal Neurons. J Neurosci vol 27, pp 8999–9008.

Larkum, M. E., Zhu, J. J., and Sakmann, B. (2001) Dendritic mechanisms underlying the coupling of the dendritic with the axonal action potential initiation zone of adult rat layer 5 pyramidal neurons. J. Physiol. vol 533, pp 447–466

Lashley, K. S. (1930) Basic neural mechanisms in behavior. Psychol. Rev. vol 37, pp 1–24

Legéndy, C. R. (1967) On the scheme by which the human brain stores information. Math. Biosci. vol 1, pp 555–597

Legéndy, C. R. (1970) The brain and its information trapping device. In: J. Rose (Ed.), Progress in Cybernetics Vol. 1 (New York: Gordon and Breach), pp 309–338.

Legéndy, C. R. (1975) Three principles of brain function and structure. Int. J. Neurosci. vol 6, pp 237–254

Legéndy, C. R. (1978) Cortical columns and the tendency of neighboring neurons to act similarly. Brain Res. vol 158, pp 89–105

Legéndy, C. R. and Salcman, M. (1985) Bursts and recurrences of bursts in spike trains of spontaneously active striate cortex neurons. J. Neurophysiol. vol 53, pp 926–939

Lettvin, J. Y., Maturana, H. R., McCulloch, W. S., and Pitts, W. H. (1959) What the frog's eye tells the frog's brain. Proc. Inst. Radio Engr. vol 47, pp 1940–1951

Livingstone, M. S. and Hubel, D. H. (1982) Thalamic inputs to cytochrome oxidase-rich regions in monkey visual cortex. Proc. Natl. Acad. Sci. USA vol 79, pp 6098–6101

Li, Z. (1998) A neural model of contour integration in the primary visual cortex. Neural Computation. vol 10, pp 903–940

Lorente de Nó, R. (1938) In: J. Fulton (Ed.), Physiology of the Nervous System (London: Oxford University Press) pp 291–340

Losonczy, A. and Magee, J. C. (2006) Integrative properties of radial oblique dendrites in hippocampal CA1 pyramidal neurons. Neuron vol 50, pp 291–307

Losonczy, A., Makara, J. K., and Magee, J. C. (2008) Compartmentalized dendritic plasticity and input feature storage in neurons. Nature vol 452, pp 436–441

Magee, J. C. and Johnston, D. (1997) A synaptically controlled, associative signal for Hebbian plasticity in hippocampal neurons. Science vol 275, pp 209–213.

Mainen, Z. F. and Sejnowski, T. J. (1995) Reliability of spike timing in neocortical neurons. Science vol 268, pp 1503–1506

Markram, H., Lubke, J., Frotscher, M., Roth, A., and Sakmann, B. (1997) Physiology and anatomy of synaptic connections between thick tufted pyramidal neurones in the developing rat neocortex. J. Physiol. (Lond.) vol 500, pp 409–440

Martinez-Conde, S., Macknik, S. L., and Hubel, D. H. (2004) The role of fixational eye movements in visual perception. Nat. Rev. vol 5, pp 229–240

McCulloch, W. S. and Pitts, W. (1943) A logical calculus of ideas immanent in nervous activity. Bull. Math. Biophys. vol 5, pp 115–133

Müller, J. (1834, 1840) Handbuch der Physiologie des Menschen für Vorlesungen von Dr. Johannes Müller (Verlag von J. Hölscher, Coblenz) 2 vols.

Muller, D., Mendez, P., De Roo, M., Klauser, P., Steen, S., and Poglia L. (2008) Role of NCAM in Spine Dynamics and Synaptogenesis. (Springer Netherlands, 2008)

Nevian, T., Larkum, M. E., Polsky, A., and Schiller, J. (2007) Properties of basal dendrites of layer 5 pyramidal neurons: A direct patch-clamp recording study. Nat. Neurosci. vol 10, pp 206–214

Noda, H. and Adey, W. R. (1970) Firing of neuron pairs in cat association cortex during sleep and wakefulness. J. Neurophysiol. vol 33, pp 672–684

Oshima, T. (1969) Studies of pyramidal tract cells. In: H. H. Jasper, A. A. Ward, and A. Pope (Eds.), Basic Mechanisms of the Epilepsies. (Boston: Little, Brown, and Co), pp 747–753.

Palm, G. (1981) Evidence, information and surprise. Biol. Cybernetics vol 42, pp 57–68

Palm, G. (1982) Neural Assemblies. (Berlin: Springer-Verlag)

Perkel, D. H. and Bullock, T. H. (1969) In: F. O. Schmitt, T. Melnechuk, G. C. Quarton, and G. Adelman, Neuronsciences Research Symposium Summaries. (MA: Cambridge and London: MIT Press), pp 405–527.

Perkel, D. H., Gerstein, G. L., and Moore, G. P. (1967) Neuronal spike trains and stochastic point processes. II Simultaneous spike trains. Biophysical J. vol 7, pp 419–440

Piper, M. and Holt, C. (2004) RNA translation in axons. Annu. Rev. Cell Dev. Biol. vol 20, pp 505–523

Purves, D. and LaMantia, A. (1993) Development of blobs in the visual cortex of macaques. J. Comp. Neurol. vol 334, pp 169–175

Rall, W. (1962) Theory of physiological properties of dendrites. Ann. N.Y. Acad. Sci. vol 96, pp 1071–1092

Ramón y Cajal, S. (1911) Histologie du Systeme Nerveux de l'Homme et des Vertébrés. (Paris: Maloine)

Rapoport, A. (1952) "Ignition" phenomena in random nets. Bull. Math. Biophys. vol 14, pp 35–44

Reich D. S., Mechler F., and Victor J. D. (2001) Independent and redundant information in nearby cortical neurons. Science vol 294, pp 2566–2568.

Remy, S. and Spruston, N. (2007) Dendritic spikes induce single-burst long-term potentiation. Proc. Nat. Acad. Sci. vol 104, pp 17192–17197.

Riggs, L. A. and Ratliff, F. (1952) The effects of counteracting the normal movements of the eye. J. Opt. Soc. Am. vol 42, pp 872–873.

Schmidt, E. M., Bak, M. J., Hambrecht, F. T., Kufta, C. V., O'Rourke, D. K., and Vallabhanath, P. (1996) Feasibility of a visual prosthesis for the blind based on intracortical microstimulation of the visual cortex. Brain vol 119, pp 507–522.

Schmidt, E. M., Jost, R. G., and Davis, K. K. (1975) Examination of the force relationship of cortical cell discharge patterns with conditioned wrist movements. Brain Res. vol 83, pp 213–223.

Schmidt, K. E. and Löwel, S. (2002) Long-range intrinsic connections in cat primary visual cortex. In: Peters, A. and Payne, B. R. (Eds.), The Cat Primary Visual Cortex. (San Diego: Academic Press) pp 387–426.

Schüz, A. and Palm, G. (1989) Density of neurons and synapses in the cerebral cortex of the mouse. J. Comp. Neurol. vol 286, pp 442–455.

Scott, A. (2002) Neuroscience: A Mathematical Primer. (Heidelberg: Springer-Verlag)

Seriès, P., Lorenceau, J., and Frégnac, Y. (2003) The "silent" surround of V1 receptive fields: Theory and experiments. J. Physiol. (Paris) vol 97, pp 453–474

Shadlen, M. N. and Newsome, W. T. (1998) The variable discharge of cortical neurons: Implications for connectivity, computation, and information coding. J. Neurosci. vol 18, pp 3870–3896

Shannon, C. E. (1948) A mathematical theory of communication. Bell System Tech. J. vol 27, pp 379–423, 623–656

Shaw, G. L. and Palm, G. (1988) Brain Theory. (Singapore, Hong Kong: World Scientific)

Sheinberg, D. L. and Logothetis, N. K. (2001) Noticing familiar objects in real world scenes: The role of temporal cortical neurons in natural vision. J. Neurosci. vol 21, pp 1340–1350

Shkolnik-Yarros, E. G. (1961) Some forms of interneuronal connections in the visual system. J. Higher Nervous Activity (Russian) vol 11, pp 680–689

Softky, W. R. and Koch, C. (1993) The highly irregular firing of cortical cells is inconsistent with temporal integration of random EPSPs. J. Neurosci. vol 13, pp 334–350

Song, S., Miller, K. D., and Abbott, L. F. (2000) Competitive Hebbian learning through spike-timing-dependent synaptic plasticity. Nat. Neurosci. vol 3, pp 919–926

Sperry R. W. (1944) Optic nerve regeneration with return of vision in anurans. J Neurophysiol vol 7, pp 57–69

Starr, P. A., Rau, G. M., Davis, V., Marks, W. J., Ostrem, J. L., Simmons, D., Lindsey, N., and Turner, R. S. (2005) Spontaneous pallidal neuronal activity in human dystonia: Comparison with Parkinson's disease and normal macaque. J. Neurophysiol. vol 93, pp 3165–3176

Stepanyants, A., Hirsch, J. A., Martinez, L. M., Kisvarday, Z. F., Ferecsko, A. S., and Chklovskii, D. B. (2008) Local potential connectivity in cat primary visual cortex. Cerebral Cortex vol 18, pp 13–28

Stevens, C. F., and Zador, A. M., (1998) Input synchrony and the irregular firing of cortical neurons. Nat. Neurosci. vol 1, pp 210–217

Stettler, D. D., Das, A., Bennett, J., and Gilbert, C. D. (2002) Lateral connectivity and contextual interactions in macaque primary visual cortex. Neuron vol 36, pp 739–750

Stettler, D. D., Yamahachi, H., Li, W., Denk, W., and Gilbert, C. D. (2006) Axons and synaptic boutons are highly dynamic in adult visual cortex. Neuron vol 49, pp 877–887

Szentágothai, J. (1973) Synaptology of the visual cortex. In: R. Jung (Ed.), Handbook of Sensory Physiology, vol VII/3B – Central Processing of Visual Information. (New York: Springer-Verlag) pp 269–324

Tomasek, J. and Hay, E. D. (1984) Analysis of the role of microfilaments and microtubules in acquisition of bipolarity and elongation of fibroblasts in hydrated collagen gels. J Cell Biol vol 99, pp 536–549

Ts'o D. Y., Gilbert C. D., and Wiesel T. N. (1986) Relationships between horizontal interactions and functional architecture in cat striate cortex as revealed by cross-correlation analysis. J. Neurosci. vol 6, pp 1160–1170

Valverde, F. (1971) Short axon neuronal subsystems in the visual cortex of the monkey. Int. J. Neurosci. vol 1, pp 181–197

Vic-D'Azyr, F. (1786) Traité d'Anatomie et de Physiologie; vol 1, Anatomie et Physiologie du Cerveau; vol 2, Planches Anatomiques. (Paris: Didot).

von Bonin, G. (1960), Jules Gabriel Francois Baillarger. In: G. von Bonin (Ed.), Some Papers on the Cerebral Cortex. (Springfield: Charles C. Thomas)

von der Malsburg, C. (1973) Self-organization of orientation sensitive cells in the striate cortex. Kybernetik vol 14, pp 85–100

von der Malsburg, C. (1999) The what and why of binding: The modeler's perspective. Neuron, vol 24, pp 95–104

von der Malsburg, C. (1981). The correlation theory of brain function. Internal Report 81–2, MPI Biophysical Chemistry. Reprinted in Models of Neural Networks II (1994), E. Domany, J. L. van Hemmen, and K. Schulten (Eds.) (Berlin: Springer).

von Economo, C. and Koskinas, G. N. (1925) The Cytoarchitectonics of the Adult Human Cortex. (Vienna and Berlin: Julius Springer Verlag) von Ehrenfels, C. (1890) Über Gestaltqualitäten. Vierteljahresschr. für Philosophie vol 14, pp 249–292

von Neumann, J. (1956) In: C. E. Shannon and J. McCarthy (Eds.), Automata Studies. (Princeton: Princeton University Press), pp 43–98.

Wertheimer, M. (1923) Untersuchen zur Lehre von der Gestalt, II. Psychol. Forschung vol 4, pp 301–305

Westheimer, G. (1976) Diffraction theory and visual hyperacuity. Am. J. Opt. Physiol. Optics vol 53, pp 362–364

Wickelgren W. A. (1999). Webs, cell assemblies, and chunking in neural nets: Introduction. Can. J. Exp. Psych. vol 53, pp 118–131

Wilson, H. R. and Cowan, J. D. (1972) Excitatory and inhibitory interactions in localized populations of model neurons. Biophysical J. vol 12, pp 1–25

Index

Printed in the United States
142974LV00003B/24/P